RUSSIAN MATHEMATICS EDUCATION

History and World Significance

SERIES ON MATHEMATICS EDUCATION

Series Editors: Mogens Niss *(Roskilde University, Denmark)*
Lee Peng Yee *(Nanyang Technological University, Singapore)*
Jeremy Kilpatrick *(University of Georgia, USA)*

Published

Vol. 1 How Chinese Learn Mathematics
Perspectives from Insiders
Edited by: L. Fan, N.-Y. Wong, J. Cai and S. Li

Vol. 2 Mathematics Education
The Singapore Journey
Edited by: K. Y. Wong, P. Y. Lee, B. Kaur, P. Y. Foong and S. F. Ng

Series on Mathematics Education Vol. 4

RUSSIAN MATHEMATICS EDUCATION

History and World Significance

Edited by

Alexander Karp
Bruce R. Vogeli
Columbia University, USA

World Scientific

NEW JERSEY • LONDON • SINGAPORE • BEIJING • SHANGHAI • HONG KONG • TAIPEI • CHENNAI

Published by

World Scientific Publishing Co. Pte. Ltd.
5 Toh Tuck Link, Singapore 596224
USA office: 27 Warren Street, Suite 401-402, Hackensack, NJ 07601
UK office: 57 Shelton Street, Covent Garden, London WC2H 9HE

British Library Cataloguing-in-Publication Data
A catalogue record for this book is available from the British Library.

Series on Mathematics Education — Vol. 4
RUSSIAN MATHEMATICS EDUCATION
History and World Significance

ISBN-13 978-981-4277-05-1
ISBN-10 981-4277-05-3

Typeset by Stallion Press
Email: enquiries@stallionpress.com

Printed in Singapore by Mainland Press Pte Ltd

Contents

Introduction

The subject of this book is mathematics education in Russia. There has been no shortage of writing on Russian (Soviet) education in general and its mathematics component in particular. Interest in this subject reached its high point in the years of the Cold War, when a period of intense scrutiny and even imitation of the Soviet school system followed in the wake of the Sputnik launch. With the advent of Gorbachev's reforms in the mid-1980s, as the country became more open, hundreds of highly qualified mathematicians poured out of Russia into the West. The prominent French mathematician Pierre Cartier remarked in jest that they had accomplished what Stalin could not do with all his army: they conquered the world (Senechal, 1998). Understandably, the educational system that had trained these mathematicians attracted particular interest, albeit sometimes a narrow one, directed toward special institutions for the mathematically talented students (a fully deserved interest, in our opinion).

As far as we know, no attempt, however, has been made, pre-Gorbachev or since, to give a systematic description and analysis of the origin and development of mathematics education in Russia. This two-volume work is an attempt to provide this description.

Volume 1 *History and World Significance* is followed by Volume 2 *Programs and Practices.* The division indicates at either instance, the dominant theme of the analysis; however, the first volume cannot avoid discussion of programs and practices, just as the second must also include a great deal of history. This entanglement is among the challenges awaiting anyone who would attempt an analysis of mathematics education in Russia. Its practice cannot be understood distinctly from its history, nor can its history be considered without an understanding of the practice.

Writing about impressions of Russia recorded by foreign travelers in the 19th century, the great culture historian Yury Lotman observes that these texts require deciphering. Offering perfectly accurate representations of time and place, they nevertheless exhibit an ignorance of a *code* that could give meaning and significance to their observations — a code rooted in an understanding of a complex and polysemic foreign culture (Lotman, 1992). The task set before the editors of this collection first of all has been to help the reader in accessing the historical and cultural codes underlying the facts in hand.

The history of the development of mathematics education is inseparable from the history of the country itself. Major efforts to bring Russia into communion with European mathematical culture were made in the time of Peter the Great (1672–1725). In the words of the poet Pushkin, the Czar had "opened a window into Europe." The contemporary cultural historian Boris Uspensky notes ironically that Peter also had raised the very wall in which this window were hacked out by bolstering serfdom, ramping up state oppression, and generally lowering literacy levels (Uspensky, 2004).

The vast country peopled with many nations for centuries had remained illiterate, even as it put forth magnificent cultural contributions exemplified by the works of Dostoevsky and Tolstoy, Mussorgsky and Chaikovsky, and Lobachevsky and Chebyshev. The revolutions of 1917 that finally handed power to the Bolsheviks appeared at first to have brought land to the peasants and rights to the formerly oppressed nations; but in the Soviet empire that rose upon the ruins of its predecessor, both rights and land were promptly expropriated. Education in general, and mathematics education specifically, had become available to an incomparably greater portion of the population, however.

Thoughts about contradictions within the social structure compelled Lev Tolstoy, along with hundreds of others among "the educated classes," to consider the possibilities of teaching mathematics to the children of peasants. Likewise, the aspiration to overcome these contradictions became the chief impetus for the students themselves. The same or similar issues and contradictions continue to fuel debate to this very day and keep present scholars from passing an unbiased judgment on the past. Old contradictions are not always resolved. Tensions of a century past are felt today as keenly as ever.

Debates over the history of mathematics education continue, and the editors of these volumes had no intention of sidestepping the issue: the reader will find in these volumes a variety of opinions and interpretations of the past as well as of the present. The authors represented here are Russian mathematics educators, who had both studied the history of the development of mathematics education and played an active role in its practice, as well as mathematics educators from other countries, closely involved in the study of the research and the practice of Russian mathematics education.

The volume opens with Tatiana Polyakova's brief history of the development of mathematics education in Russia prior to 1917. The following chapter (Alexander Karp) discusses the formation of the Soviet system of mathematics education in the first half of the 20th century. Alexander Abramov continues the thread with a discussion of the history of the so-called "Kolmogorov reform" of the 1960s–1980s. Mark Bashmakov offers an analysis of the recent past of Russian mathematics education and submits his vision of the key challenges facing the system today. The chapter written by Alexey Sossinsky addresses a purely Russian phenomenon — the extensive involvement of research mathematicians in the school mathematics education. Mark Saul and Dmitri Fomin focus on a particular manifestation of this involvement, mathematics competitions, tracing their history, and illustrating it with diverse examples of problems typically offered to participants. Jean Schmittau discusses the program of mathematics teaching in elementary schools, devised by the prominent psychologist Vasily Davydov that has gained great popularity abroad. Natalya Stefanova talks about the history and the practice of pre-service mathematics teacher education. The reader will also find in this chapter useful information on the organization of mathematics training in Russian middle and high schools today. The next chapter partitioned into three sections, written by Antoni Pardała, Katalin Fried, and Orlando Alonso, examines the influence of Russian (Soviet) mathematics education in three countries of the former so-called socialist block (i.e., countries falling within the sphere of influence of the

USSR). The countries selected are Poland, Hungary, and Cuba. Finally, Jeremy Kilpatrick writes about the influence of Soviet psychological studies in the USA and about the history of their publication in English.

The second volume offers a detailed discussion of the practice of mathematics education in Russia, focusing separately on the distinct disciplines (i.e., algebra, geometry, calculus, and finite mathematics), the specifics of the curricula in schools for mathematically talented students, schools with reduced mathematics components, and elementary schools, as well as on mathematics-related extracurricular activities. Special attention will be given to lesson planning in Russian schools and the main trends in research in mathematics education.

Finally, we note that several of the chapters included in this volume were originally written in Russian and subsequently translated into English. The editors wish to thank Ilya Bernstein and Sergey Levchin and also Heidi Reich for help in preparing the manuscript for publication.

References

Lotman, Yu. M. (1992). K voprosu ob istochnikovedcheskom znachenii vyskazyvanii inostrantsev o Rossii [Toward the problem of historical significance of foreign pronouncements on Russia]. In Yu. M. Lotman, *Izbrannye stat'i v trekh tomakh*, Vol. 2 (pp. 138–144). Moscow: Aleksandra.

Senechal, M. (1998). The continuing silence of Bourbaki. An Interview with Pierre Cartier, June 18, 1997. *The Mathematical Intelligencer*, *1*, 22–28.

Uspensky, B. A. (2004). Evropa kak metafora i kak metonimiya (primenitel'no k istorii Rossii) [Europe as metaphor and metonymy (as Applied to Russian History)]. In B. A. Uspensky, *Istoriko-filologicheskie ocherki* (pp. 9–26). Moscow: Yazyki slavianskoy kul'tury.

1

Mathematics Education in Russia before the 1917 Revolution

Tatiana Polyakova

Pedagogical Institute
Southern Federal University, Russia

1 The History of the Inception of Russian Mathematics Education

The period from the 10th to the 17th centuries can be called the Age of the Inception of Mathematics Teaching in what is now Russia. It was during this period that conditions gradually changed, permitting mathematics education to emerge in the 18th century as a broad national concern. The historical records of this period lack specific evidence of the content or methods of mathematics education and most of the individuals who contributed to the emergence of mathematics education remain unknown. Still, historical evidence of commerce, government, and military activities indicates that mathematical activity was ongoing and was transferred by some means from person to person and generation to generation. Development proceeded in a haphazard manner, however; so the following account describes only the vital stages in the development of mathematical education in Russia.

The first stage took place in Kievan Rus', which in the 10th–12th centuries reached its zenith in terms of both culture and sheer power. Byzantium was a principal influence for both "intellectual and literary activity" in Kievan Rus' (Kostomarov, 1995, p. 9); it brought the

Cyrillic alphabet to Russia, which stimulated the development of a unified system of letters and numbers.

Prince Vladimir, who brought Christianity to Rus', and his son, Yaroslav the Wise, were the first to realize the importance of education. The Orthodox church expected educated people to support the newly accepted religion. On the strength of this belief public schools were founded in Kiev, in Novgorod and in other prominent cities, primarily for the children of priests and the secular upper classes, but not for the rest of the population. Education was mandated aggressively by the state. The result was the first few generations of educated people in Russia.

There are a few literary sources documenting the quality of education at that time. The best are the juridical collection "Russian Truth" by Yaroslav the Wise (Grekov, 1947) and the first mathematical essay in Rus' by the monk Kirik (Istoriko-matematicheskie issledovaniya, 1952).

"Pravda Russkaya" (Russian Truth) contains many articles with mathematical calculations. Some of these are quite complicated for that time (computation of percentages, evaluation of areas, increases in livestock, and other chattel). The collection's wide circulation implies that the educated segment of society could both understand and use its mathematical laws.

The very highest level of mathematical scholarship in Rus' is demonstrated in the mathematical–chronological essay "The manual of how a person comes to know numeration of years" (Nastavlenie, kak cheloveku poznat' schilslenie let) written by Kirik Novgorodets in the beginning of the 12th century. It contains professional-level mathematical–chronological calculations and even an example of a geometric progression with a common ratio of five.

At this time, however, mathematics education was most likely subordinate to other forms of education and had only a utilitarian intention. Its contents were limited to elementary information from practical geometry and also the rudiments of arithmetic. In the 10th–12th centuries in Rus', among the elite, in any case, there was an acceptance of the value of education and the quality of mathematical education fully comparable with Byzantine and European models.

The second stage (13th–14th centuries) coincides with the Tatar-Mongol invasion. The general cultural decline of this period included a decline in all levels of education at all levels of society. For all intents and purposes, schools virtually ceased their existence. Even the most educated societal group, the clergy, experienced this decline. The chronic insufficiency of literate people was so severe that even positions as priests were unfilled. Additionally in the 15th century, the clergy became the savage enemy of the dissemination of mathematics, all but banning mathematical books.

The only city that retained high culture in Old Rus' was Novgorod, which was virtually untouched by the Tatar-Mongol invasion. Many groups of society there had strikingly high levels of education, as evidenced by the sensational archeological discovery of writings on birch bark in Novgorod in the middle of the 20th century. The writings had sufficiently many numerical figures, including some written by children, to testify to the high level of mathematical education. This discovery, which evidenced the first appearance of the so-called "numerical alphabet," marks the first educational material in mathematics; the birch bark writings were presumably used for the study of numeration and exercises in the writing of numbers (Simonov, 1974, p. 80).

The third stage is linked with the consolidation of power in Moscovian Rus' (15th–17th centuries). The church's prohibition notwithstanding, literature appeared in which there was mathematical material. Geometry and arithmetic were apparently used in practical activities and art (Polyakova, 1977, p. 33). The clergy was the first to embrace the value of education, and monasteries played an important role in the development of enlightenment. As a result, libraries and schools were created in conjunction with monasteries.

The ecclesiastical system of education grew. In 1639, the first establishment of higher education was opened: the Kiev-Mogilyansky Academy. "In Russia, as in other countries, the need for higher education was satisfied earlier than the need for middle or lower education" (Brockhaus and Efron, 1898, p. 382). In the final year of the Academy, the curriculum included elements of geometry. Moscow's Slavic–Greek–Latin Academy, which was opened later, did not include mathematics in its curriculum.

In the 17th century, mathematical education functioned on a high level, even outside the clerical educational system, as evidenced by a considerable quantity of mathematical manuscripts from the period. The overwhelming majority of these manuscripts appear to be educational textbooks in mathematics. In turn, arithmetic manuscripts constituted the majority of these textbooks. The quality of the arithmetic textbooks matches that of contemporary European prototypes (Yushkevich, 1968, p. 24). Their contents included numeration, rules of operations with whole numbers and fractions, calculation, rules of commercial arithmetic, and elements of entertaining arithmetic (puzzles). These arithmetic manuscripts were largely responsible for the dissemination in Rus' of the Indo–Arabic system of numeration, which was of paramount importance for the general cultural development of the country.

Methodologically speaking, arithmetic was approached dogmatically. As in other countries, artificial rules were applied (the rule of three, the rules of regula falsi, etc.) to problem solving. The Russian abacus appeared as the basic calculating instrument. Arithmetic manuscripts used the traditional Russian system of measurement. Problems often were rooted in specific Russian reality (Polyakova, 1977, pp. 60–61).

The matter of geometry was different. Geometry was usually included in arithmetic manuscripts and a few practical manuals providing information for solving practical problems; the rules in these manuscripts were often inexact and occasionally incorrect and their foundations were absent. Only two manuscripts of that period were dedicated entirely to geometry. One was a textbook of practical geometry containing reasonably reliable rules of measuring distance and area. It also included problems on construction and isometric transformations of figures. The level of geometric mastery in this book is quite low, markedly lower than that of its European contemporaries (Polyakova, 1977, p. 67).

However, this is not the case with the second geometric manuscript. "Sinodal'naya No. 42" (Belyj and Shvetsov, 1959) occupies a prominent place in the history of mathematical education. For the first time in Russia this geometry textbook, created by order of the sovereign

Mikhail Fyodorovich, contains a systematic account of geometry presented in a manner similar to its European counterparts. It contains definitions of geometric figures, theorems with diagrams and elements of proofs, and solutions of problems on construction and calculation. The textbook was not printed nor widely distributed, and although it did not influence the development of mathematical education in Rus', it was an indicator of an existence of a layer of educated people in 17th century Rus', who not only were interested in mathematics but were also interested in its dissemination (Polyakova, 1977, p. 74).

In the 17th century, for the first time since the time of the Kievan Rus', the idea of the value of education caught the attention of the country's leadership. Political will, however, proved insufficient to cause real educational change. Tsar Boris Godunov intended to open schools and even universities, but he was unsuccessful; Mikhail Romanov commissioned a modern geometric textbook but did not facilitate its publication. Finally Peter I, often referred to as Peter the Great, had the kind of political will necessary to bring about real change.

2 The 18th Century: The Period in which Mathematical Education in Russia Came to a Halt

2.1 *Mathematical Education in the Epoch of Peter I*

Once the governmental reform began, Peter I was thwarted by the absence of literate people who were adequately prepared to bring his plans to fruition. For this reason, he began the preparation of specialists to form a regular army, to build a fleet and to open factories, and to reconstruct the apparatus of government. Peter's direct participation brought about the first secular public schools; moreover, he established a clear dominant position for mathematics in every secular school. In doing so, Peter I set the precedent for governmental patronage of mathematical education (Polyakova, 2000, p. 175).

From the beginning remembering his own European journeys, Peter I tried to utilize the scientific-educational potential of Europe.

Before Peter's rule, only diplomats and merchants were authorized to cross the border. Under Peter, however, travel abroad were encouraged and even mandated. As it turned out, Peter's initiative was largely unsuccessful. Only a few young Russians aspired to study abroad; even fewer proved capable.

The lack of books, necessary for the dissemination of knowledge, was another obstacle. To remedy the situation, in 1700 Peter I gave Yan Tessing, a businessman from Amsterdam, the right to publish and sell books in Russian secular stores. Losses were suffered: there were not very many booklovers in Russia and the quality of books was not high. All the same, the first academic books appeared in Russian, and they were mathematical. Despite this limited success of book production, however, by and large, attempts to utilize the scientific-educational potential of Europe did not produce the desired effect.

In the beginning of the 18th century, Peter I embarked on the first organization in Russia of a national, secular, public, and professional educational system. Peter I was sufficiently competent in mathematics to appreciate its role in military-technical education. For this reason, mathematics was one of the fundamental subjects in organized schools, the teaching of which he and his comrades-in-arms followed personally.

In 1701, mathematical–navigational and artillery schools were opened in Moscow. In 1707, a surgical school was opened, affiliated with a military hospital. In 1711–1712, an engineering school was opened. Subsequently a few training colleges appeared in conjunction with factories in Karelia and the Urals, where metallurgical craftsmen were trained.

2.1.1 *The mathematical–navigational school*

On 14 January 1701, Peter I issued a decree for the foundation of a school in Moscow "for mathematical and navigational, that is, seafaring scientific skills." It produced young people "for all sorts of service, military and civil, that demanded scientific knowledge or even Russian language-based knowledge; from the navigational school there emerged, besides sailors: engineers, artillerists, teachers for other new schools, geodesists, architects, civil servants, clerks, craftsmen and others" (Veselago, 1852, p. 7). The school was intended for children of gentry and civil servants. However, the gentry and civil servants were not overly anxious to educate their children; besides, they did not want their children to study with others of "ignoble birth." As a result a significant number of the school's students were from a lower class of society. The school enrolled a total of 500 students from 12 to 17 years of age and was housed in Sukharev Tower. Sukharev Tower acquired a wide reputation and was considered a major center in both mathematics and general scholarship.

During the time the Tsar was in England, he became acquainted with a professor from Aberdeen University named Fargwarson and invited him and his two colleagues to Russia. There, Fargwarson became one of the founders of the mathematical–navigational school where he established a thriving program. He took part in the development of the educational system and brought in programs in arithmetic, algebra, geometry, plane, and spherical trigonometry; he himself taught and also wrote the textbooks. But the English people did not speak Russian and the students not only did not speak English, they frequently could not read or write Russian also. Subsequently Leontij Filippovich Magnitsky, whom Peter I both knew and respected, was invited to accept a teaching post. Magnitsky was one of the most educated people of his time. He knew Latin, Greek, German, and Dutch and was acquainted with the achievements of European mathematicians. He taught arithmetic, geometry, and trigonometry, fulfilling his duties with exceptional conscientiousness. Magnitsky became the senior teacher and head of the academic part of the school. He was the leader of the mathematical–navigational school for the remainder of his life.

2.1.1.1 Magnitsky's "Arithmetika"

Several handbooks in mathematics were written by L. F. Magnitsky, of which the most important, "Arithmetika," was printed in Moscow in January 1703. Twenty four hundred copies were printed, which was a large number for that time. In the next half-century it was popular in schools, but enjoyed even wider popularity among other readers, particularly autodidacts. This was largely due to its linking of traditional, Moscow-based educational literature with new European influences. Since Magnitsky knew foreign languages so well, he was able to master a large quantity of European textbooks, books by Greek and Latin authors, and manuscripts by Russian mathematicians. He incorporated all of these materials in his textbooks.

The textbook's name conveys some but not all of the book's contents. It also introduced significant algebraic and geometric material and elements of plane and spherical trigonometry in addition to arithmetical knowledge. Because of its breadth, "Arithmetika" functions as an encyclopedia of mathematical knowledge of the time.

In accordance with the tradition of Russian educational literature, Magnitsky included the "Deed of Peter" in "Arithmetika." It functions on some level as a textbook of the latest Russian history (although an apologetic variant thereof). Also, it had many general philosophical debates and advice to the reader (frequently in verse). "Arithmetika" contained information on meteorology, astronomy, and navigation, as well as numerous data on natural science and engineering. "Arithmetika" was a precursor of popular scientific literature in addition to its other merits.

2.1.1.2 Organization of instruction

Normative measures by today's standards, such as lesson plans and programs, were not found in the mathematical–navigational school. The content of education was defined on the basis of L. F. Magnitsky's "Arithmetika" and on the geometry textbook "Priyomy tsirkul'a i linejki" ("Using compass and straightedge") and "Geometriya praktika" ("Geometry for the practitioner") which were translated into Russian by Ya. V. Brius. Instruction was dogmatic: it demanded that students memorize rules and be able to apply them to problem solving. At the same time, Magnitsky's "Arithmetika," which defined the method of instruction, was not devoid of methodical merit: its examples were selected with increasing difficulty and presented an interesting array of problems.

Study at that time was not easy. Classes were taught in a poorly-understood language, instructional equipment was scarce, and instructors interacted with students sternly, often using corporal punishment. For these reasons students frequently left the school.

In 1715, navigational classes were moved to Petersburg where the Morskaya (Navy) Academy was founded. At that time, Russia was a formidable naval power. With the Academy's opening came the reconstruction of the curriculum: war-related science was studied in the Academy. In the Moscow school, they studied mathematics only, which prepared the students for the Academy's course.

Difficulties notwithstanding, the mathematical–navigational school secured a prominent role not only in Russian mathematical educational history, but also in Russian education as a whole. It provided rapid production of high-quality mathematicians and navigators, as well as specialists of wider description. It became the first center for promoting secular schooling at that time, first and foremost for mathematics. It also became the first institution of teacher training, preparing mathematics teachers for a large number of learning institutions.

2.1.2 *Arithmetic (Tsifirnye) schools*

Arithmetic (Tsifirnye) schools (schools devoted to teaching counting and arithmetic in general) were established in 1714 in provincial cities in conjunction with bishoprics and large monasteries. Their student composition was relatively heterogeneous, including children of nobles and civil servants, those in the priesthood, and merchants. "Education was free, but at the conclusion of study, before issuing a certificate, a teacher had the right to collect a ruble for each student. Without this certificate it was forbidden to marry" (Kostomarov, 1995, p. 351). As mentioned above, the mathematical–navigational school prepared the first teachers for the arithmetic schools: Peter I ordered that two students from the navigational school well-versed in geometry and geography were to be sent to each of the provinces to teach there.

In 1716, 12 arithmetic schools were opened; in 1720–1722, another 30. A few more than 2000 people were conscripted into these schools, both voluntarily and by force. Educational reforms had formerly met with opposition from society. "Arithmetic study" was posted as mandatory for upper and middle social classes, but gradually children of nobility, townspeople, and clergy were freed from mandatory attendance from these schools, and subsequently 14 of the 42 existing schools were closed. The remaining students were almost exclusively children of civil servants. Toward 1727 only 500 students remained. In 1744, arithmetic schools were merged with garrison schools, opened in 1716 by Peter's decree for the education of soldiers' children.

2.1.2.1 Mathematical education in arithmetic schools

Students in arithmetic schools studied arithmetic and geometry. There were no established textbooks. It appears that the only source of mathematical knowledge was the teacher. Frequently one teacher conducted lessons for 20–30 students who all were studying totally different subjects at the same time — the equivalent of the American one-room schools. There were neither demonstrations nor intelligible explanations. The teacher would formulate basic definitions and rules and would provide solutions for model problems. The students were required to memorize a series of rules and solve problems.

2.1.2.2 Difficulties in teaching

About 15% of students left the arithmetic school, although it was not uncommon for them to be put in prison, in chains, for doing so (Gnedenko, 1946, p. 51). The causes for this were diverse. The first group of causes was linked with the opposition of society. As previously discussed, the population reluctantly allowed their children to go to school because it sharply changed old family structures, customs, and habits. The school regime was very cruel; poor results and carelessness were treated with corporal punishment. The second group of causes was related to the fact that the system of instruction was still "a work in progress." Textbooks, tablets, and instruments were in short supply. Teachers had no special preparation. There was no teaching methodology. Russian educational terminology, including a terminology of mathematics, was poorly cultivated. Despite relatively low efficacy, a few thousand Russian students were educated in State schools in Peter's time.

2.2 *Leonhard Euler and Mathematical Education in Russia*

2.2.1 *The academic educational system*

Earlier mathematical education in Peter's time was characterized as practical. Leonhard Euler took part in the foundation of the educational system that may be called "academic" (Polyakova, 1977, p. 127). It arose linked with the 1724 establishment of the Imperial Petersburg Academy of Sciences; this system functioned for the entire 18th century and ceased at the beginning of the 19th century.

Peter I decided to strengthen the Academy not only scientifically, but also in its teaching capacity, thereby providing the economic basis for the university and *academic gymnasium* (Gnedenko, 1946, p. 71). This was an original idea, as was the affiliation of the university with the Academy. In contrast to European universities, Peter's university consisted only of juridical, medical, and philosophical departments, with the exclusion of a theological department. The Academy, university, and gymnasium were therefore wholly secular institutions.

Still another distinction of the Petersburg Academy from European academies was that it was not public, but rather an organ of the government, supported firmly by a governmental budget. In fact, one of Peter I's final decrees set a tradition of governmental patronage of science. Since the caliber of invited scientist-mathematicians was excellent, so the development of mathematics as a science was strongly effected and bolstered.

After Peter I's death, the idea of the value of education to some extent lost importance in the eyes of the country's highest leadership. Educational politics caused external support to languish, but for the most part the material stimuli for the development of education continued due to inertia (Polyakova, 2000, pp. 177–178). Moreover, up until that time different levels of society considered the Academy (and its educational foundations) a very important institution and so it enjoyed some self-sufficiency and was able to function. It was important for the extension of the governmental patronage that the Academy focused not only on research but also on instruction.

The appearance of Leonhard Euler was extremely fortunate for Russian mathematics and mathematics education. As Peter I is considered responsible for a vigorous organizational influence on Russian education, so Euler imparted a great strength of content and methodology, creating and incorporating the mechanism of research mathematics' patronage of mathematical education. This patronage is fully demonstrated in the activity of the Mathematical–Methodological School of Leonhard Euler, which is a unique phenomenon in Russian intellectual history.

2.2.2 *The mathematical–methodological school of Leonhard Euler*

In the second quarter of the 18th century, the informal mathematical–methodological school, founded by Euler, played an ever-increasing role in the development of mathematical education. Though it is easy to identify the time when this school began its work — it coincided with the beginning of Euler's activities — it is a bit difficult to identify its end, largely because the ideas of Euler's methodological school developed over the course of the entire 18th century, and in many respects continued to be significant in the 19th century. Mathematics education as a scientific field was born later; therefore, the usage of the term "mathematical–methodological school" itself might have been put into question. Nevertheless, understanding the whole evolution of this term, we accept this term here.

The first Russian academicians-mathematicians S. K. Kotel'nikov, S. Y. Rumovskij, and M. E. Golovin, and the academician-secretary of the academy, N. I. Fuss, assembled the skeleton of the Euler school (Lankov, 1951, pp. 15–18). It is considered to be the first methodological school in Russia, since Magnitsky was a one-man methodologist. Moreover, although one of his students, N. G. Kurganov, developed the methodological ideas of Magnitsky, he also developed those of Euler, and factually belongs more to the methodological school of the latter. The school's activity affected the development of mathematical teaching in Russia in the following substantial ways:

- First, it provided operational access to the pedagogical and methodological ideas of Europe, in which the idea of proof and the systematization of the exposition of mathematics dominated.
- Second, having acquired these ideas, the methodologists of the school enriched and made sense of them. Adapting the ideas to Russian reality facilitated their quick introduction.
- Third, they fairly swiftly put non-translated, original Russian educational mathematical literature into use instead of foreign sources.

Euler arrived at the Petersburg Academy of Sciences in 1727. In the course of his service in addition to intensive scientific activity, he occupied himself with the teaching of mathematics in the academic educational system. He attended to the selection of the content of mathematical education, having had experience in writing several mathematics textbooks specifically for the academic gymnasium. Among them, "Manual in arithmetic for use in the gymnasium of the Imperial Academy of Sciences" bears mention. It was published in German (1738–1740) and in Russian translation (1740, 1760). This book had significant influence on academic mathematical literature, standing as a precedent for the foundation of prominent academically-accessible school textbooks written at a high scientific level.

Later Euler wrote his algebra textbook, which, again, stood as a prototype for all subsequent textbooks. Euler published "Full introduction to algebra" in Petersburg in German in 1770. In 1768–1769, a Russian translation was published named "Universal arithmetic." This book contained material that would have been sufficient even for a university course, but some sections of this book were used successfully in the academic gymnasium.

There is reason to suppose that Euler also wrote a geometry textbook. There is reference to it in a number of bibliographic sources and a few manuscript fragments have been found (Belyj, 1961, p. 186). Euler also developed textbook materials in modern trigonometry, presented in almost the same form as is studied today in schools. Thus,

Euler had textbooks in practically all mathematical academic subjects which boasted the most modern methodical ideas of the time.

The first such idea was the reconciliation of mathematical educational content with modern mathematics. It found its embodiment in an algebra textbook by Euler, subsequently reworked by N. I. Fuss; in trigonometry textbooks by M. E. Golovin and S. Y. Rumovskij; and in a textbook of mathematical analysis by S. K. Kotel'nikov. For the first time in academic courses, the newest achievements of mathematicians were included such as Euler's modern trigonometry, and his work in differential and integral calculus. (Everything referenced here and further textbooks of Euler are characterized in detail in Polyakova, 1977, pp. 143–157, 184–198.)

The second important methodological idea realized by Euler was the introduction of fundamental mathematical disciplines in school mathematics education (i.e., arithmetic, geometry, trigonometry, and algebra) as separate and specific subjects. This approach helped to discourage unnecessary diversity in previously-used mathematics textbooks (for example, a popular in Europe textbook by Christian Wolff included the study of 19 disciplines, all of which were considered mathematical). Euler's approach helped to clear textbooks of materials foreign to the course of each school's particular discipline. In this way, the arithmetic textbooks of Euler and Kurganov were cleansed of elements of algebra and geometry.

The third methodological idea realized by Euler in his textbooks is the building of mathematical courses on the basis of progressive (for that time, but even for today) didactic principles including systematicity, scientific foundation, and accessibility to the students of the exposition of mathematical discipline. Importantly, Euler speaks not only about these principles but (in modern language) about finding the optimal combination of a high-level scientific foundation and accessibility by the student.

Representatives of Euler's methodological school, with few exceptions (N. G. Kurganov, M. E. Golovin), were academicians — the intellectual elite of society. Having organized mathematics instruction in all types of schools and the creation of textbooks, they actively participated in scientific-organizational (Kotel'nikov,

Rumovskij, Fuss), educational-cultural and popularizing (Kurganov, Kotel'nikov, Rumovskij, Fuss), and instructional-organizational (Kurganov, Kotel'nikov, Rumovskij, Golovin) activities. Euler and Kotel'nikov, in addition, authored original studies on the reorganization of high school education.

2.3 *Mathematical Education in Russia in the Second Half of the 18th Century*

In the second half of the 18th century the most advanced levels of Russian society began to realize the value of education. There were several educational systems in Russia at that time: professional, academic (described above), university, public schools, and schools affiliated with the church (which will not be covered here).

2.3.1 *The professional educational system*

The professional educational system was reconstructed during this period to serve increasingly the children of the upper class and the nobility. The system included military schools (both the navy and the army "noble-born" cadet corps), military-technical schools (engineer–artillery noble-born cadet corps), technical (mining) schools, and others. Towards the middle of the 18th century the largest viable educational field was mathematics. Mathematics was considered to be a field of high priority and was distinguished by its high quality.

2.3.1.1 The naval cadet corps

The teaching of mathematics in the naval cadet corps was particularly good. Teachers in these schools were particularly well-prepared. The most famous pedagogue-mathematicians of Euler's methodological school worked in naval corps. Professor N. G. Kurganov devoted his entire career to the education of future marine officers. The academician S. K. Kotel'nikov combined teaching in the academic gymnasium with delivering lectures to those in the naval corps and to writing textbooks for their usage. The academician N. I. Fuss wrote several mathematics textbooks and taught a course not only in the navy corps, but also in the army cadet corps.

In the 1760s the navy corps began to offer to the upper class a course on higher mathematics in addition to the elementary mathematics course. The founder of this tradition was S. K. Kotel'nikov, who started teaching sections of higher mathematics especially for naval cadets. In 1766, his textbook "First Foundations of Mathematical Science" ("Pervye osnovaniya matematicheskikh nauk") was published. At the end of the 18th century, the course in mathematics was supplemented with analytic geometry and mathematical analysis and was taught by N. I. Fuss.

2.3.1.2 The engineering–artillery corps

The director of the *engineering–artillery* cadet corps M. I. Mordvinov, believed that mathematics should take precedence in officer preparation. He included mechanics, arithmetic, the beginnings of algebra, and higher geometry in the syllabus. Mordvinov was concerned with teacher quality, primarily in mathematics.[1] One of the very best teachers there was Y. P. Kozel'skij, well-known for his philosophical essays. Kozel'skij had attended the academic gymnasium and the university where he was a student of the famous Russian scholar Lomonosov (after whom Moscow State University was named later). It is interesting to note Kozel'skij's opinions on instruction as expressed in the prefaces to his textbooks and philosophical works where he emphasized the link between theory and practice. In 1764, Kozel'skij published an arithmetic textbook, "Arithmetic propositions" ("Arifmeticheskie predlozheniya"), which was distinguished by its clarity and concreteness with particular attention to material that would be needed in everyday life.

---

[1]The wages of the faculty were determined by the importance of their subject and the availability of the teachers. It is interesting to note the following list of wages. Mathematics teacher: 800 rubles per year; political science: 600 rubles; Russian language: 500; foreign language: 400; and dance teacher: 300 rubles per year.

N. V. Vereshchagin, a disciple of Kozel'skij, was one of the best educated teachers in the corps. He had extensive knowledge of mathematics and military disciplines, and was also extraordinarily knowledgeable about natural science, philosophy, and history. He knew French, German, Italian and Latin, which gave him the ability to follow developments in mathematics and to reflect in his lectures upon the latest mathematical achievements. Vereshchagin familiarized his cadets with Euler's *Introduction to Infinitesimal Analysis* and also taught them, among other things, methods of solving systems of linear equations with determinants. He was one of the first people in Russia to teach analytic geometry. Vereshchagin set an example in Russia for disinterested service. He gave lectures to students and amateurs who had an inclination toward mathematics without any compensation. In addition, he exerted substantial influence not only on mathematical education in the engineering–artillery corps but also on the development of mathematical education in general in the second half of the 18th century.

2.3.2 *The university educational system*

The most important element of the system was *Moscow University.* With the opening of the Naval Academy in Petersburg (1715), the center of mathematical education moved from Moscow to Petersburg. After the foundation of the Academy of Sciences, Petersburg became the definitive focal point of the development of mathematics and of mathematical education. New military and technical schools were opened, followed by a teaching seminary. Petersburg was the center of Euler's scientific-methodological school; it was the site where he published his works about mathematics and its teaching.

The founding of Moscow University in 1755 renewed Moscow's perspective concerning the development of mathematics and mathematical education. The university, however, did not have a mathematics department; one could study mathematics only as an ancillary subject. In the course of almost half a century, only minimal amounts of mathematics necessary for medicine and natural science were included in the Moscow University curriculum. These courses also were taken by students who wished to be considered generally well-educated.

A specialized chair of mathematics was established in 1758 at Moscow University at the request of the medical faculty. Professors of medicine noticed that students had deficient knowledge of physics because they were unable to *apply* their mathematical knowledge (since the only course offered in mathematics at the time was pure mathematics). The first specialist-mathematician to occupy the chair of mathematics was D. S. Anichkov, who was the head of the mathematics faculty and who had published textbooks in many mathematical subjects. Anichkov taught a course in pure mathematics in a two-year cycle, with two two-hour lectures per week. In the first year, he taught arithmetic and geometry; in the second, a continuation of geometry and trigonometry. Surprisingly, algebra was not always offered. After Anichkov the mathematician V. K. Arshenevskij taught with the same time arrangement per week, but his course had a three-year cycle which included algebra in the third year. I. A. Rost and his successor, M. I. Pankevich, taught applied mathematics in a three-year cycle of four two-hour lectures per week.

The level of mathematics teaching was lower than in academic educational establishments and professional schools of the time. The principal merit of university mathematics education was the active preparation of prospective teachers. At the end of the 1760s there were so called "gymnasia informers" (*Biographical dictionary*, p. 46), students prepared for future pedagogical work, to whom special courses were given. After a decade a pedagogical seminary was built, which, beginning in the 1790s, was called the "Teachers' Seminary." The mathematical education curriculum there included arithmetic, geometry, planar trigonometry, and elements of algebra. The Teachers' Seminary prepared teachers to teach in Moscow University's gymnasia, and eventually for gymnasia teaching in general.

"With a university there should be a gymnasia," wrote M. V. Lomonosov, the founder of Moscow University, "without which the university is like a plowed field without seeds" (Lomonosov, 1957, p. 514). In 1755, two gymnasia were founded for children of nobility and for students of other societal levels. Shortly after 1755, the university opened a gymnasium in Kazan'. These gymnasia were successful because less-affluent nobles were happy to educate their children there instead of hiring expensive (and frequently poorly-prepared) foreign teachers. In 1779, a boarding school was founded for the nobility. Based on this model, schools for "children of noble birth" were created in Nizhnij-Novgorod, Tver', Ryazan', Kursk, and other provincial cities. In this fashion, Moscow University became a leading supporter of school education since the university supplied teachers, textbooks, and occasionally money.

Two gymnasia in Moscow shared a single rector. Instruction was carried out separately in each school until the last class (the so-called "Rector class"), for which the students from both schools, often including children of lower rank or children of impoverished nobles, were brought together for preparation to enter the university. The children of the affluent nobility on completion of their first classes typically prepared for military or civil service rather than attending the Rector class. Once enrolled in the University, student "received a sword and with it a noble rank. Upon finishing university, the student left with the rank of an officer" (Brockhaus and Efron, 1898, p. 383). The provision of an officer's rank inspired the lower-ranking students to obtain a university education, thereby gaining not only intellectual advancement, but also material resources and prestige.

Although generally education at that time emphasized the humanities, there was not just one educational trajectory for all gymnasia students. Gymnasia students could select trajectories of training themselves. Technically, the gymnasium was divided into three schools: Russian, Latin, and the school of new European languages. The Russian school was an introductory one. After graduating the student could choose between the Latin school (leading to the University) and the school of new languages. Mathematics was not taught in the Russian school. Later, students had a course in arithmetic (twice a week for 4 hours). In the next class they studied fractions, elements of geometry, and algebra (4 hours per week). In the "highest" class they continued the study of arithmetic, algebra, and geometry for 4 hours per week. In the beginning of the 1790s, algebra and trigonometry were added to the gymnasia's curricula.

The general level of mathematical education in the university educational system, including gymnasia in large cities and in Moscow, was extremely low. There was one sole exception, and that was Moscow University's boarding school for children of noble birth. The focus in this school was "practical" studies rather than humanities.

2.3.3 *The general education system: Public schools at the end of the 18th century*

In the first half of her reign, Katherine II was very interested in questions of upbringing, following the ideas of the French encyclopedists. She attempted to realize various educational projects. The organization of networks of homogeneous schools of general education which covered the entire country proved of particular importance for Russian education, including mathematical education. In 1782, the "Commission to establish public schools" began its work on a "plan to establish public schools in the Russian empire." The Commission followed an Austrian educational model and invited an Austrian Serb, Yankovich de Mirievo, a follower of the ideas of Jan Amos Komensky, to carry out the reforms. De Mirievo knew Russian and had experience with school organization gained during the foundation of a system of training colleges for Austrian Serbs living in Hungary. In 1786, in accordance with "Rules of education for the people," schools were opened in the cities in Russia, including so-called "chief schools" in the capital of provinces (guberniya) and so-called "small schools" in the major towns of districts (uezd).

For this project to be successful, it was essential to prepare teachers, and so, in 1782, a special "teachers' seminary" was opened. Students were recruited from Russia's religious schools — the best students from the Alexander Nevsky, Smolensk, and Kazan' orthodox seminaries as well as students from the Slavic–Greek–Latin Academy were invited to study there. Yankovich tried to acquaint students with new methods of teaching, outlined in his "Handbook for Teachers of First and Second Classes of the Public Schools" ("Rukovodstvo uchitel'am pervovo i vtorovo klassov narodnykh uchilishch"), 1783. This book was the first methodological handbook in the history of Russian education in which requirements for a method of teaching were found. In this way, a foundation for the organization of the

Russian educational system and its characteristic methodology was prepared.

Mathematical subjects were among the most important in the curriculum, according to the rules of the schools. In the small schools, students studied arithmetic: in the first class, they would do oral and written calculations; in the second, there was a continuation of arithmetic including the rule of three. In the first and second classes in the chief schools, students would review the program of the small schools (to avoid any omissions); in the third class, the faculty taught fractions, decimals, and also exercises in the rule of three and others. In the fourth class, students studied basic geometry in an abbreviated form.

In his "Handbook for Teachers..." Yankovich proposed a method of teaching oral and written calculation. Teachers would have to clarify each new rule with examples and detailed commentary. After that, they would give an assignment with that rule to the best student who would complete it on the classroom board and then comment on its solution. Then the teacher would give an assignment that the class would solve on their individual boards, or one student would record the solution on the classroom board, prompted by the class. Later, when basic skills were acquired, the students were given tasks based on different rules without instructions and were asked to determine which of them should be used for the solution. These activities were expected to instill appropriate applications of different rules. The questions selected by the teacher were not to be abstract, but rather related to life. At home, students were supposed to solve new problems that were always checked during the next lesson.

Textbooks available for the public schools were not attributed to specific authors. Later, this practice proved to be a source of discussion among educational historians. The majority of historians attribute the authorship (in full or in part) to M. E. Golovin, who belonged to the methodological school of Euler. The organization of these textbooks was quite traditional. In the first part of the "Handbook for Arithmetic," for example, numeration and operations were presented with whole and nominative numbers; in the second part, fractions, decimals, proportions, and "exercises with rules" were discussed. The geometry textbook contained only 16 theorems with proofs. The bulk of the material consisted of fairly interesting problems. Substantial attention was given to measurement, for which the usage of physical models was recommended. In the summer, Golovin offered what we would now call "lesson-excursions."

3 The Formation of Russia's Classical System of School-Based Mathematical Education: 19th–early 20th Centuries

3.1 *Mathematical Education in Russia in the First Quarter of the 19th Century*

Following the foundation of the Ministry of People's Education in 1802, Russia was divided into six academic regions, with the leading role in each entrusted to that region's university. In 1803, "four types of schools were identified: 1) parish-based schools, 2) district-based schools, 3) provincial schools (called gymnasia), and 4) universities" (Yushkevich, 1948, p. 15). In the parish-based schools, instruction lasted one year, in the district schools two, and in the gymnasia four. Continuity was provided between them, although that continuity deteriorated almost completely toward the beginning of the 20th century. In 1804, the rules of the universities and other academic establishments were accepted, which consolidated the universities' scientific-methodical role, providing in this way the most qualified possible leadership of early and secondary education.

The value of education at that time was accepted by both the higher leadership of the country and a wide portion of society. Educational politics of the leadership oscillated depending on exterior and internal circumstances: from clearly-declared democratic tendencies at the beginning of the century to the idea of classicism, toward the end of the first quarter of the century, which was considered as a counterbalance of democratic tendencies.

Changes occurred also in the school's class-based restrictions. At the beginning of the 19th century, almost all class-based restrictions were removed. The government took over the financial support of the gymnasium. Toward the beginning of the 1820s, regulations became more rigid; in particular, people now were required to pay for their education.

During this period in Russian education, the opposition between the classicist approach (based on the study of ancient languages) and the "real" approach (natural-mathematical) became pronounced. This opposition was particularly acute at critical moments in Russian history. Its influence upon the quality of mathematical education was indicated by:

(1) the number of students enrolled,
(2) the position of mathematics in a general system of education.

In the beginning of the century, gymnasia education, on the whole, included aspects of practical ("real") education. Mathematics occupied the primary position in the schedule of lessons. Six hours per week were allotted to the study of mathematics during the first three (out of four) years. During the last two years of education, students studied statistics as well (4 and 2 hours per week, respectively). Mathematics teachers (as opposed to teachers of language or drawing) were called senior teachers and occupied the highest rank of governmental civil service a teacher could have, that is the ninth rank (according to Peter I's ranks system). A teacher of mathematics held a rank equal to the military rank of captain. A mathematics teacher's elevated rank allowed mathematics to rank the highest place among the academic subjects. Moreover, the government officially proclaimed education free of class and cost. In the beginning of the century, the government provided, practically speaking, the ideal conditions for the development of education in general and mathematical education in particular.

Its content at this time was determined neither by the preparation nor interests of the teacher, as it had been at the beginning of the 18th century, nor by mathematical textbooks, as it was the end of the century, but by rules of the academic institution. In the parish schools the first operations of arithmetic were studied; in the district schools, arithmetic and beginning geometry were presented; in the gymnasia "pure and applied mathematics and experimental physics" were part of the curriculum. The names of the subjects themselves speak to the fact that the model of mathematical education characterized by the 18th century had not yet been overcome.

Pure mathematics included geometry, plane trigonometry, and algebra up until third degree equations with applications to geometry and conic sections. Courses in pure and applied mathematics included some sections of physics, and also elements of analytic and descriptive geometry, with the beginnings of differential and integral calculus. In the first year of gymnasia, students studied pure mathematics consisting of algebra, geometry, and plane trigonometry; in the second year, the study of pure mathematics was completed and applied mathematics and experimental physics were introduced, to be concluded in the third year.

The leading influence on the content of mathematics education, as before, was the content of mathematics textbooks. For the improvement of old textbooks and to prepare new ones, the Academic committee was appointed in 1803. One of the appointed members was academic-mathematician N. I. Fuss. Beginning in 1805, the first two volumes of "Course in Mathematics" by T. F. Osipovsky was recommended as the textbook for gymnasia. The first volume contained arithmetic and algebra; the second, geometry, rectilinear, and spherical trigonometry and an introduction to curvilinear geometry. The textbook "Course in Mathematics" presented rich content with simplicity and accessibility. It was very popular in secondary and higher educational institutions and was reprinted many times. This success notwithstanding, this textbook was not actually intended for gymnasia, but as a university textbook in mathematics. The Academic Committee decided to put together textbooks specifically for gymnasia, the preparation of which was commissioned to N. I. Fuss.

N. I. Fuss reworked earlier publications and published "Beginning Foundations of Pure Mathematics," which from 1814 until 1828 functioned as the first stable textbook of mathematics for gymnasia in Russia. Fuss included the foundations of algebra and geometry, applications of algebra to geometry, plane trigonometry, conic sections, and the foundations of differential and integral calculus.

The tradition of support of mathematical education by research mathematics continued in this period. This tradition of patronage was evidenced by the work of practically all of the eminent mathematicians. In particular, S. Y. Rumovsky and N. I. Fuss were the first members of the Academic Committee. S. E. Gur'yev published the first mathematical–methodological essays in Russia (Polyakova, 2002, pp. 219–237). T. F. Osipovsky, the future rector of Khar'kov University and one of the directors of the academic district, provided important assistance to mathematical education on all levels.

3.2 *The Development of the Gymnasia System of Mathematical Education in the Second Quarter of the 19th Century*

Although there was an attempt to bring uniformity to gymnasia education, the quest for uniformity was not successful. In fact, this very question had been part of the discussion of the reform of the educational system in Russia over many years. In 1826, the "Committee to organize academic institutions" ("Komitet ustrojstva uchebnykh zavedenij") was established in order to bring about this reform: by establishing new regulations for academic institutions. Gymnasia regulations, established in 1828, announced a seven-year period of study. Three years of elementary school were included in the gymnasia course. Strict class-based boundary restrictions were established: only children of nobility and merchants of the first guild were allowed to enroll in gymnasia. The cost of studies was increased; the classical underpinnings were emphasized. These "innovations" had chiefly a negative effect on mathematical education. The only positive effect of the 1828 rules was the elimination of vagueness of topics included in mathematical courses. Applied mathematics was eliminated. Pure mathematics was limited to cover only topics through conic sections (additionally, the course of descriptive geometry was introduced).

There was also a bifurcation in gymnasia education. It was divided into two types: the first, or Variant A, included the study of Latin. The second, or Variant B, included both Latin and Greek. It could be surmised that the first variant embodied a practical approach, whereas the second a classical one.

The content of mathematical education in these types of gymnasia was diverse. In the regulations of 1828, besides the schedule of lessons, a plan of study was included for the first time in the history of Russian mathematical education. This showed not only academic discipline, but also contained subjects of study that would provide consistency. In gymnasia of the first type, students studied arithmetic in the first and second classes; in the third, the beginnings of algebra were studied, including second degree equations; in the fourth, students finished algebra and started geometry; in the fifth, they finished geometry; in the sixth, they studied elements of descriptive geometry and the beginning of applications of algebra to geometry; in the seventh, they completed applications of algebra to geometry up to conic sections. At the completion of the course, the teacher conducted a review of the whole course of mathematics. In the second type of gymnasia, the course in mathematics was shortened to 15 hours per week (combined for all grades)[2] and included only arithmetic in the first and second classes, algebra in the third, and geometry in the fourth, fifth, and sixth classes, and a review of these offerings in the final year of studies.

Despite deficiencies, the regulation of 1828 played a positive role in the development of Russian mathematical education by providing firm limits on its content. The implementation of these changes shows that society as a whole was beginning to appreciate the general cultural meaning of mathematical education.

The first concise unified programs in mathematics introduced in 1832 were intended to guarantee uniform mathematics instruction in gymnasia across the entire country. It succeeded in providing relative uniformity only on the level of the academic regions. Each region's trustees delegated the task of developing precise instructions on mathematics teaching in the gymnasia to the regions' universities.

---

[2] By comparison, 46 hours per week (combined for all grades) were allotted to the study of Greek and Latin. (Here and below, we frequently indicate the combined number of hours per week for all grades, i.e., if during the first four years of study, 3 hours per week were allotted to some subject, and during the subsequent three years, 2 hours per week were allotted to the same subject, the combined total for all seven grades would be 18 hours per week.)

From the end of the 1820s until the beginning of the 1830s, there was a delicate equilibrium between the classical and "real" (practical) models of gymnasia education. This bifurcation was bolstered by the regional interpretations of the national regulations for the gymnasia.

The position changed significantly in 1834 with the appointment of S. S. Uvarov to the post of Minister of Education. He was active in educational politics, central to which was the famous triad "Orthodoxy, Autocracy, Nationality." The political remedy chosen to comply with this triad was to tighten class boundaries and raise the price of education. The academic plan was revised to exclude topics such as statistics and logic. The dominant model for gymnasia education was the classical system with its basis in ancient languages.

As a result of the Uvarov "reforms," mathematics, in large part, lost its former privileged status. In 1845, a circular governmental letter appeared called "On the shortening of the education of mathematics in gymnasia" ("Ob ogranicheniyakh v gimnaziyakh prepodavaniya matematiki"), in which the teaching of analytic and descriptive geometry in the gymnasia was abolished. The usual number of hours per week in mathematics combined for all four classes was 20 hours. A new allocation of mathematical classes was proposed in 1846. Based on this allocation, F. I. Busse developed his project of an exemplary program in mathematics. In it he broadened the course of arithmetic and algebra and extended the time allotted for this course. Trigonometry was introduced in Busse's project in an effort to reinforce the applied side of the mathematical disciplines.

The 1848 anti-monarchist revolutions in Europe brought new changes to educational politics in Russia. The Ministry of Education published a memorandum intended to eradicate "free thinking" in high schools and universities. This time, the conflict between classical and practical ("real") gymnasia education was settled in favor of the latter. Instead of Greek, students were required to study natural-mathematical disciplines. At the same time, the number of hours per week devoted to mathematics combined for all four classes was increased from 20 to 30, however, in 1852 that number was reduced to 22.5 hours.

Once again a new program of mathematics was introduced but it was not significantly different from the program of 1846. Special attention was given to practical applications of mathematics and the strengthening of interdisciplinary links between the sub-fields of mathematics.

The textbook, "Beginning foundations of pure mathematics" ("Nachal'nye osnovaniya chistoj matematiki") by N. I. Fuss, was replaced by arithmetic textbooks and exercise books written by F. I. Busse and "Arithmetic leaflets" ("Arifmeticheskie listki") by P. S. Gur'yev. A translation of "Course in pure mathematics" ("Kurs chistoj matematiki") by the French mathematician Bellavein also was used in Russian gymnasia.

"Handbook for arithmetic" ("Rukovodstvo k arifmetike") and "A collection of arithmetic problems for gymnasia and district schools" ("Sobranie arifmeticheskikh zadach dlya gimnazij i uezdnykh uchilishch") by F. I. Busse were the textbooks of a new type (Polyakova, 2002, p. 277). For the first time, the complete set of books for students and teachers united by one methodological approach was created. This set included a textbook, an exercise book, and a teacher guide entitled "Handbook for instruction in arithmetic" ("Rukovodstvo k prepodavaniyu arifmetiki"). This handbook was the first methodological handbook for Russian mathematics teachers. Complete sets of textbooks, problem books, and teacher guides did not become customary until the end of the 20th century, suggesting that the methodological provision of F. I. Busse's arithmetic course can be considered a "breakthrough" in the history of mathematical education. F. I. Busse's textbook appeared in a record number of 18 editions from 1830 to 1875. It is worth noting also that Busse's collection of problems was one of the first such collections published in Russia, establishing a tradition continued to the present.

P. S. Gur'yev published "Arithmetic leaflets" with the primary goal of developing students' self-sufficiency and assisting teachers in providing differentiated instruction. Their original form utilized individualized sheets with portions of learning materials to be distributed to students in accordance with their ability.

"A course in pure mathematics" by Bellavein was a reworked and supplemented variant of a course that was quite full, but compact. The most successful part of this text was on algebra, which was published separately and accepted officially for usage in gymnasia until the 1850s.

Despite efforts to provide uniformity, various academic districts used different textbooks, or occasionally local manuscript versions, that reflected the preferences of the region. Thus, in Moscow gymnasia textbooks by D. M. Perevoshchikov were used; in Kazan', manuscripts and textbooks by N. I. Lobachevsky were preferred.

The textbook "A gymnasia course in pure mathematics" by D. M. Perevoshchikov included practically the whole university course in mathematics — arithmetic, elementary algebra, beginning geometry, rectilinear trigonometry, and conic sections. In the gymnasia, courses in algebra and trigonometry were offered widely. The textbook "The foundations of algebra" was published in 1854, and was recommended by the Ministry as a textbook for gymnasia. Excerpts in trigonometry from Perevoshchikov's prior text were recommended for gymnasia as late as the 1860s.

Textbooks in elementary geometry and elementary algebra by N. I. Lobachevsky (1823, 1825), which appeared as manuscript, were used in gymnasia in the Kazan' academic district. These manuscripts generalized the gymnasia course and were used in gymnasia in an abbreviated form as an introduction to university study. Lobachevsky's geometry textbook was developed along the lines of the author's original methodological system, the basis of which is laid out in the methodological essay "A survey of teaching pure mathematics." Beginning with the measurement of circumference, Lobachevsky used the theory of limits. He utilized the idea of the parallel study of plane and three-dimensional geometry. This essay defines the methodical innovation of N. I. Lobachevsky.

N. I. Lobachevsky's textbook on elementary algebra was not published at that time. In 1835, only his "Algebra, or calculation of finites" was printed, in which topics such as the foundations of operations with whole numbers, a general approach to functions, an original treatment of transcendental functions, and other topics were presented with great self-sufficiency and methodical innovation.

In 1844, textbooks developed especially for gymnasia were published. In geometry, "Mastery of beginning geometry" by F. I. Busse and the exercise book "Practical exercises in geometry" by P. Gur'yev and A. Dmitriev appeared. In arithmetic, the text "Arithmetic" by V. Y. Bunyakovsky appeared.

Busse's textbook was the first geometry textbook in the history of Russian mathematical education with modern structure and contents. It did not contain a separate treatment of straight line geometry, including only two- and three-dimensional geometry. The Busse textbook corresponded to the program of 1846 and was well designed methodologically. The exercise book included problems devoted to constructions and calculation, arranged in increasing order of difficulty.

V. Y. Bunyakovsky's "Arithmetic" appeared in three editions (1844, 1849, 1852) and was used in the majority of academic regions and enjoyed deserved popularity due to its recognized methodological merit. To its credit, it had systematic and comprehensive examples as well as complete and understandable explanation of rules. It was well organized with a limited amount of basic materials and with all supplementary materials moved to the appendix. All irrelevant traditional materials were excluded. There were methodical notes for readers, concentrated primarily in the preface and then generalized subsequently in "Program and abstract of arithmetic."

The tradition of the research mathematicians' patronage of mathematics was strengthened by the universities' dominant role in academic–methodological leadership in the schools. V. Y. Bunyakovsky was a prominent mathematician of this period who wrote textbooks for the schools, taking part in methodological meetings and commissions. Bunyakovsky taught in military-academic institutions, and even published one of the first books on the methods of teaching mathematics. The great mathematician, N. I. Lobachevsky, who was mostly unrecognized by his contemporaries, fulfilled his obligation as one of the leaders of the Kazan' academic region and wrote textbooks in elementary geometry and elementary algebra. He also wrote serious methodical works and even published a pedagogical essay (the published speech "On the most important subject in upbringing").

We should take special notice of the role in the development of mathematical education of mathematician–university rectors D. M. Perevoshchikov and N. I. Lobachevsky. They stood at the head of the Moscow and Kazan' academic regions, providing competent academic–methodological leadership in schools that put school-based mathematical education in their regions on a high level. In addition, they wrote first-class mathematical textbooks.

3.3 *Mathematical Education in Russian Schools in the Second Half of the 19th Century and the Beginning of the 20th Century*

In 1858, an Academic Committee appointed by the Ministry of Education issued new school regulations, foreseeing the bifurcation of gymnasia education into philological and physical–mathematical specializations. The prominent mathematician and founder of the Petersburg mathematical school, P. L. Chebyshev, guided the creation of the academic plan in mathematics. His plan was formulated with the following goals:

(1) the development of the intellectual capability of students;
(2) the mastery of knowledge vital to any cultured person;
(3) preparation for specialized occupations in physical–mathematical sciences with attention to their practical application.

For this plan, the gymnasia course in mathematics was partitioned into two parts: "general," for all students in the 1st–5th classes; and "specialized," for those wishing to continue their education in the "practical" vein in the 6th–8th classes. The specialized course included continuation of algebra, trigonometry, analytical and descriptive geometry, mathematical geography, optics, and mechanics. The right to develop a specific syllabus was granted to the faculty committee of each gymnasium. However, this project did not come to fruition. The Academic Committee rejected the idea of bifurcation in 1859.

A new project of gymnasia regulations was proposed in 1860 in which "real" (more practical) education was favored. Latin was begun only in the third class, Greek in the fifth. The number of hours per week allotted to mathematics was 27.5 hours (combined for all years of the gymnasia studies). Twenty hours per week were allocated for natural science and physics. This project was altered twice, in 1862 and 1864, finally resulting in a decrease of hours allocated to mathematics.

The problems of education, including mathematical education, were discussed widely and publicly in the 1860s in pedagogical and public-political periodicals. A professional discussion of the problems of gymnasia education in general took place in 1864 at a pedagogical conference for school directors and teachers in Odessa. In particular, at the Odessa meeting, the issue of gymnasia mathematical education was discussed. Concerns about content, in terms of the overload of the program of 1852, were resolved by a decision to abbreviate the course in mathematics. The conference formulated methodological principles in the teaching of mathematics, such as striving for consciousness in learning and visualization; in addition, teachers' intervisitations with a full exchange of ideas were recommended.

New regulations for gymnasia and programs in accordance with the foundation of three types of gymnasia with seven-year terms of study were established in 1864. These regulations recognized three types of curricula: "classical" with two ancient languages, classical with one ancient language, and "real" or applied curriculum. The regulations provided the opportunity to choose among the three, granting equal rights to graduates.[3] Corporal punishment also was abolished while the authority of teachers' and faculty committees was elevated.

P. L. Chebyshev revised the pedagogical guidelines in 1865, sharply defining the framework of the teaching of mathematics, physics, and cosmography. In the program for the classical gymnasia 22 hours per week were allotted to mathematics; in the real gymnasia, the allocation for mathematics was 25 hours per week. The content of the course in mathematics was defined by general characteristics only. The teachers had the right to design their syllabi with subsequent approval by faculty committees. This allowed the teaching of mathematics to distinguish itself not only in gymnasia in different regions, but also in a single region.

---

[3]All of the real gymnasia, however, guaranteed admission only to higher specialized academic institutions and, for the first time, to physical–mathematical university departments. The classical schools did not have an analogous limitation.

In 1871, the liberal school regulations of 1864 were replaced by regulations which remained in place without substantial change until 1917. All of the gymnasia were transformed into classical gymnasia with an eight-year term of study including two ancient languages. Mathematical study in gymnasia was abstract-deductive, and formal scholastic instruction methods were encouraged. Mathematics, physics, mathematical geography, and natural science together were allotted just 37 hours out of the 206 hours per week. By comparison, Latin was allotted 49 hours and Greek was allotted 36 hours.

"Real schools" were established to replace "real gymnasia." After the sixth class, students in real schools selected one of three directions: commercial, mechanical–technical or chemical. Real school did not provide the study of ancient languages. However, the course in mathematics was substantial. Further, many hours were allotted to technical drawing. Upon completion of the real school, students were denied the right of admission to a university. The battle between classical and real education, which went on for the entire 19th century, concluded with a victory for the classical model of school education. The quantity of hours devoted to mathematics in school did not decrease, however, but even increased. In addition, Russian schools had, for the first time, stable national curricula in all subjects, including mathematics.

Published in 1872, this syllabus in mathematics, which owed its existence to P. L. Chebyshev, defined the following order of mathematics study in gymnasia. In the first through third classes, students studied arithmetic. In the third class they began the study of algebra, which was completed in the seventh year of education. In the eighth year students repeated the entire courses of algebra and arithmetic. The geometry course began in the fourth class and was completed in the seventh year with a review of everything that had been studied and with the study of the application of algebra to geometry. Trigonometry was studied during the eighth year.

The mathematical content remained roughly the same as that described in the 1852 syllabus. Changes were introduced with the goal of eliminating parts considered either too difficult or unimportant. Real schools' academic plans and curricula in all subjects including mathematics, mechanics, and practical mechanics were approved in 1873. New mathematics curricula for real schools were based on the academic plan prepared by P. L. Chebyshev that was briefly characterized earlier. Considerable attention was paid in the curriculum to descriptive geometry, series, and theory of limits.

New regulations for educational institutions were issued for real schools in 1888 and for gymnasia in 1891. However, they had practically no influence on the teaching of mathematics. Several not particularly substantive amendments were added to the 1872 curriculum in 1890. In 1890, for the first time in the history of Russian schools, an explanatory note on teaching methodology was included in all curricula. Also, it was emphasized that, as an exact and abstract science, mathematics should be a means of intellectual development. For this reason, basic attention should be paid to the study of theory, while practical examples should serve as an illustration of theory and for the acquisition of calculating skills. This approach substantially weakened the representation of the role of applications of mathematics.

The 1890 arithmetic curricula included the following sections: the decimal system of numeration, arithmetic operations with rational numbers, their application to the study of quantities, the metric system of measurement, and the Russian abacus. The course was completed with the section "Problems based on rules." In the next class, students studied elements of theoretical arithmetic: systems of numeration, theorems about divisibility and factoring of numbers, indicators of divisibility, repeating decimals, and approximations.

The syllabus of algebra included the following sections: polynomials, algebraic fractions, first-degree equations, the theory of proportion, equations and second-degree trinomials, radicals, quadratic and cubic roots of numbers, quadratic roots of polynomials, progressions, logarithms, equations and systems of equations, first-degree inequalities, indeterminate first-degree equations, continuous fractions, the theory of combination, and the binomial theorem. The next class included divisibility of polynomials by binomials, equivalence of equations, applications of algebra to geometry, the theory of limits, and the existence of logarithms.

The syllabus in the real schools was distinguished by the inclusion in the last class of complex numbers, solutions of binomial equations with applications to the construction of inscribed regular polygons, analysis of the extrema of some rational functions, and the method of indeterminate coefficients.

The syllabus in geometry was partitioned in two parts:

Plane geometry — the study of angles, parallel lines, triangles, polygons, circumference, incommensurate segments, similarity, inscribed and circumscribed polygons, metric properties of triangles, calculation of area, the concept of limit and the calculation of circumference and area of circles, and geometric constructions.

Solid geometry — straight lines and planes in space; polyhedra (prisms, parallelepipeds, pyramids); regular polyhedra; congruence and similarity of prisms and pyramids; cylinder, cone, sphere, and sections of these bodies.

Although the 1890 mathematical syllabus remained in Russian schools until 1917, several attempts at reform took place early in the 20th century. At some point "classicism" was somewhat weakened in the gymnasia — Greek became no longer mandatory, the number of hours in Latin was decreased, and the written examination in Latin was eliminated. The usual number of hours per week allotted to mathematics hovered around 30.

At the turn of the century, the Ministry of education decided to bring the problem of secondary education to the forum of public opinion. As a result, academic plans and syllabi were composed for four types of schools: gymnasia with two ancient languages, gymnasia with one ancient language, real school, and a new type of secondary school. The following goal of the teaching of mathematics was formulated: "mathematics would be adopted as a science in its own right and as a scientific method of the exploration." A mathematical subcommission carried out the bulk of the work (Metel'skij, 1968, p. 12) with the belief that mathematical content in the gymnasia should be identical to that in real schools.

Change in the syllabi of the real schools took place in 1906, when the academic plan of the senior classes was altered to include the beginning of analytic geometry (except for analysis of second-order curves) and mathematical analysis.

Toward the 1890s, a system took shape that can be called an international classical system of mathematical education following I. K. Andronov (1967, p. 6). Here are the particulars of its mathematical content:

(1) The study of so-called elementary mathematics in secondary school and higher mathematics in higher educational institutions.
(2) The division of elementary mathematics into four academic subjects: arithmetic, algebra, geometry, and trigonometry.
(3) The study in elementary school of arithmetic only and with only an informal empirical approach.
(4) The study in higher institutions of the bases of mathematics from the 17th and 18th centuries — analytical geometry and mathematical analysis, known as "higher mathematics."

The methodical particulars of this system were:

(1) The establishment of two goals of education: "purely educational," directed at mastery of mathematical facts; and "developmental," aimed at the development of the student's formal-logical thought.
(2) A sharp delimitation of the functions of the teacher and student: the teacher would deliver prepared lessons, the student would memorize and recall passively.
(3) The availability of textbooks and workbooks for every academic sub-division in mathematics.

This classical system of mathematical education, which endured for a very long time, had some outstanding results, although, at the same time, it had extreme inherent deficiencies:

(1) The disparity between the developing "science of mathematics" and the very stable "academic subject of mathematics."
(2) The "divide" between elementary and higher mathematics.
(3) The absence of an informal preparatory course in mathematics (in geometry, in particular).

(4) Weak connections between the four sub-divisions of elementary mathematics.
(5) The prevalence of formal-logical goals for the study of mathematics and negligence of other goals of its study.
(6) The prevalence of problems and exercises of artificial character, poorly related to practice or even theory.

As a result of this development in history, a pyramidal system emerged in Russia where only 40–50% of those who had enrolled in the first class graduated from school. This extensive lack of success in mathematics engendered an extreme point of view — to exclude mathematics completely from the general education school course.

3.4 *The Reform Movement in Russian Mathematics Education at the Turn of the Century (19th–20th)*

By the beginning of the 20th century, difficulties and clashes in school mathematical education had accumulated. As a result, a movement toward reform emerged. This movement gained an international character linked with the name of Felix Klein. In 1897, at the International Mathematical Congress in Zurich, Klein established the necessity of reform in mathematical education and formulated its principles which formed the foundation of the "Meran Programme," developed in 1905 under Klein's direction.

Even earlier, similar ideas were widely discussed in Russia. During 1891, a work called "The goal and means of the teaching of elementary mathematics from the point of view of general education" by S. I. Shokhor-Trotskij was published in the journal "*Russkaya Shkola*" (Russian School) (Nos. 2, 3, 9, 10). In this work, the existing system of mathematics education was criticized, and the introduction of a beginning course in geometry, the enrichment of arithmetic with approximating calculations, and the introduction of the study of functions and the theory of limits in the algebra course were proposed. A lot of attention was attracted by the publication of V. P. Sheremetevsky's article "Mathematics as a science and its school-based surrogates" in

1895, in which the author emphasized that modern mathematical ideas were ignored by the gymnasia curriculum. Sheremetevsky believed that the elementary course in mathematics should be based on the concept of functional relation and fought for the introduction of elements of analytical geometry and mathematical analysis into the school course. A great amount of attention was devoted to the problems of reform in mathematical education in pedagogical and methodological publications and in academic-pedagogical discussions in Petersburg and Moscow. For example, special summer courses were organized for teachers; in 1911, the journal "Matematicheskoe obrazovanie" (Mathematical education) appeared, which became a center of educational discussions. Toward the end of the 19th century, Russian mathematical education not only became a part of the international classical system of mathematical education, but was focused on development, reform, and principles which coincided with global tendencies.

The international reform movement was organized officially at the Fourth International Congress of Mathematicians held in Rome (1908). At that meeting Klein was elected president of the executive committee of the International Mathematical Commission. Well-known mathematicians including the academician N. Y. Sonin and professor K. A. Posse headed the Russian subcommittee of this Commission.

Two All-Russia congresses of mathematics teachers contributed to the advancement of the reform movement. During the winter holidays of the 1911–1912 academic year the first All-Russia congress of mathematics teachers met in Petersburg with 1217 educators in attendance. In all, 71 presentations were given, followed by discussions. The proceedings of the congress were published (Publications, 1913). The well-known mathematician, philosopher, mathematical historian, and pedagogue D. D. Morduchay-Boltovsky (1912) provided a detailed report on the conference. Among the important items discussed at the congress were the following:

(1) It was suggested that elements of analytical geometry and especially mathematical analysis were to be introduced into school "only

with the assistance of graphical, visual methods to clarify this idea and only to the extent that it can be done in that way" (Morduchay-Boltovsky, 1912, p. 9). M. G. Popruzhenko, F. B. Filippovich, and D. D. Morduchay-Boltovsky presented serious objections to this position.

(2) The reconstruction of the school course in geometry with more rigor, and the list of axioms, in particular, did not receive support; as opposed to the idea of an informal preparatory course in geometry built on a visual and intuitive approach rather than on logical reasoning.

(3) General problems and issues attracted considerable interest, including philosophical aspects of mathematics, psychological questions on the teaching of mathematics, the introduction of elements of history of mathematics, and the practical and applied directions of school mathematics.

The second All-Russia meeting of mathematics teachers took place in Moscow during the winter holidays of academic year 1913–1914 with 1200 participants. In all, 22 presentations were delivered (Reports, 1915). Seven resolutions were accepted, the first of which concerned the preparation of mathematics teachers. The following proposal was made:

(a) That people entering the teaching mathematics should possess preparation both in subject matter and in pedagogy.

(b) That in physical–mathematical departments of higher academic institutions courses should be taught that would illuminate, from an academic point of view, the foundations of aspects of elementary mathematics.

(c) That regional conferences of mathematics teachers should be established.

(d) That short-term and long-term pedagogical courses for mathematics teachers should be established.

(e) That higher academic institutions and mathematical societies should assist in organizing these courses (Reports, 1915, pp. 163–164).

The following proposals were made with the goal of improving the curriculum in mathematics:

(a) That the right be granted to gymnasium faculty committees to permit teachers to deviate from existing official syllabi with the condition that changes would be approved by schools' academic committees.
(b) That change of the syllabi and of the plan of the teaching of mathematics in school be carried out in their entirety and not in parts. It would be imperative that such a project not only would introduce new topics, but also would eliminate some topics that had become outdated and irrelevant.
(c) That the teaching of mathematics in girls' gymnasia be organized in exactly the same manner as in boys' gymnasia.
(d) That reworking of the plans and syllabi of teaching should be done in conjunction with representatives from the universities and scientific and teachers' societies (Reports, p. 164).

The Congress identified analytical geometry and analysis as essential to schools of all types, recommending:

(a) A review of syllabi of analytical geometry and analysis.
(b) The allocation of sufficient time to the study of these topics.
(c) The establishment of connections between analysis and previous parts of the course.
(d) Improvements in the teaching of analytical geometry and analysis (Reports, p. 164).

Many of the problems discussed in the Congress were not resolved. A few fruitful ideas were not put into effect due to World War I. Also a few of the ideas proved unsupportable in practice. The Congress's work showed that the Russian reform movement in mathematics education had a wide and rich content and various forms. To a great extent, the influence of the Congress was magnified by the extensive increase during the 1912–1915 period in the publication of academic and mathematical–methodological literature with new content and ideas from the Congress.

4 Conclusion

The kernel of Russian mathematical education — the Russian model of the international classical system of school-based mathematics education — was established before the 1917 Revolution. It immensely influenced everything that happened in Russian mathematical education in the 20th century. From 1918–1931, there were attempts to eradicate the classical system of mathematical education, accompanied by a generally unsuccessful search for new models. From 1931–1964, there was a restoration of Russian traditions with the formation of the Soviet version of the classical system of school-based mathematics education. This Soviet version achieved its greatest functionality in the 1940s–1950s, but practically exhausted its resources by the 1960s. From 1964–1981 there was a reformation of the Soviet model of the classical system of mathematical education (often referred to as the "Kolmogorov Reform"). In 1982, a counter-reformation movement began, which in large part returned mathematical education to the classical model.

References

Andronov, I. K. (1967). *Polveka razvitiya shkolnogo matematicheskogo obrazovaniya v SSSR* [*Fifty Years of Development in Mathematics Education in the USSR*]. Moscow: Prosveschenie.

Belyj, Y. A. (1961). Ob uchebnike Ejlera po elementarnoj geometrii [About Euler's textbook on elementary geometry]. *Istoriko-matematicheskie issledovaniya,* Pub. XIV.

Belyj, Y. A., and Shvetsov, K. I. (1959). Ob odnoj russkoj geometricheskoj rukopisi pervoj chetverti XVII v. [About one Russian geometric manuscript from the first quarter of the 17th century]. *Istoriko-matematicheskie issledovaniya.* Pub. XII.

Biographical Dictionary of Professors and Instructors of Moscow University [*Biograficheskij slovar' professorov i prepodavatelej Moskovckovo Universiteta*]. (1855). Part 2. Moscow.

Brockhaus, F. A., and Efron, I. A. (1898). *Entsiklopedicheskij slovar' "Rossiya"* [*Encyclopedic Dictionary "Russia"*]. St. Petersburg.

Gnedenko, B. V. (1946). *Ocherki po istorii matematiki v Rossii* [*Essays on the history of mathematics in Russia*]. Moscow.

Gnedenko, B. V. (1980). Moskovskij universitet i matematicheskoe prosveshchenie [Moscow university and mathematical instruction]. *Matematika v shkole.* No. 1.

Grekov, B. D. (Ed.). (1947). *Pravda Russkaya* [*Russian Truth*] M.: Pub. AN USSR.

Kostomarov, N. I. (1995). *Russkaya istoriya v zhizneopisaniakh ee glavnejshikh deyatelej* [*Russian History in Biographies of Its Greatest Personages*]: In two volumes. Vol. I. Moscow.
Lankov, A. V. (1951). *K istorii razvitiya peredovykh idej v russkoj metodike matematiki* [*Towards a history of the development of advanced ideas in Russian methodology in mathematics*]. M.: Uchpedgiz.
Lomonosov, M. V. (1957). *Polnoe sobranie sochinenij* [*Full collection of works*]. In 11 volumes. Vol. 10. Moscow.
Metel'skij, N. V. (1968). *Ocherki istorii metodiki matematiki* [*Essays on the History of Mathematical Methodology*]. Minsk: Vyshejshaya shkola.
Morduchay-Boltovsky, D. D. (1912). *Pervyj Vserossijskij s'ezd prepodavatelej matematiki* [*The First All-Russia Congress of Mathematics Teachers*]. Warsaw.
Morduchay-Boltovsky, D. D. (1914). *Vtoroj Vserossijskij s'ezd prepodavatelej matematiki* [*The Second All-Russia Congress of Mathematics Teachers*]. Warsaw.
Polyakova, T. S. (1977). *Istoriya otechestvennovo shkol'novo matematicheskovo obrazovaniya* [*The History of Russian School Mathematical Education*]. Two centuries: Book 1 Rostov/D, Pub. RGPU.
Polyakova, T. S. (2000). Otechestvenye paternalistskie traditsii matematicheskovo obrazovaniya v XVIII i pervoj polovinye XIX v. [The Russian paternalist tradition of mathematical education in the 18th century and the first half of the 19th century]. *Istoriko-matematicheskie issledovaniya*. Second series. Pub. 5(40).
Polyakova, T. S. (2002). *Istoriya matematicheskovo obrazovaniya v Rossii* [*The History of Mathematical Education in Russia*]. M.: Moscow University.
Publications of the First All-Russia Congress of Mathematics Teachers [*Trudy 1-vo Vserossijskovo S'ezda Prepodavatelej matematiki*]. (1913). St. Petersburg.
Reports that were read at the Second All-Russia Congress of Mathematics Teachers in Moscow [*Doklady, chitannye na 2-m Vserossijskom S'ezde prepodavatelej matematiki v Moskve*]. (1915). Moscow.
Simonov, R. A. (1974). "Tsifrovye alfavity" i sostoyanie gramotnosti v Drevnej Rusi ["Numerical alphabets" and the state of literacy in Old Rus']. *Matematika v shkole*, No. 1.
The manual of how a person comes to know numeration of years [Nastavlenie, kak cheloveku poznat' schilslenie l'et]. (1952). *Istoriko-mathematicheskie issledovaniya*. VI.
Veselago, F. F. (1852). *Ocherk istorii Morskovo kadetskovo korpusa* [*Essay on the History of the Navy Cadet Corpus*]. St. Petersburg.
Yushkevich, A. P. (1948). Matematika i ee prepodavanie v Rossii v XVII-XIX vv. [Mathematics and its instruction in Russia in the 17th–19th centuries]. *Matematika v shkole*, # 1.
Yushkevich, A. P. (1968). *Istoriya matematiki v Rossii* [*The History of Mathematics in Russia*]. Moscow.

2

Reforms and Counter-Reforms: Schools between 1917 and the 1950s

Alexander Karp

Teachers College, Columbia University
New York, USA

1 Introduction

This chapter deals with the formative period of the Soviet system of mathematics education. This period is not homogeneous: during its first years, immediately following the Revolution of 1917, the old, pre-Revolutionary system was deliberately and methodically destroyed with a view to construct a new revolutionary system to take its place. This was followed by an equally emphatic return to the old, but under new conditions, thus leading to the rise of a new system, although one that borrowed heavily from the pre-Revolutionary period.

Current attitudes toward what went on during these four decades vary widely. Kolyagin (2001, p. 145) views the works of reform-minded writers from 1917–1918 as models of incompetence, utopianism, and impracticality, while seeing the period following the restructuring of education by the Central Committee of the Communist Party in the 1930s as a golden age of stability. By contrast, Gurkina (2001), while acknowledging that during the 1930s "a stable school system with successive stages"[1] took shape in the country, points out its shortcomings, such as "the lack of alternatives and excessive unification of the principles underlying the contents and organization of the educational process, the refusal to make differentiations in education," and so on (although she also concedes that "such defects were partly compensated for by the efforts of talented or simply conscientious teachers" (Gurkina, 2001, p. 50)). One can also point to Western publications that in one or another way reveal a sympathy for the educational style of the 1920s, much more liberal by comparison with the period that followed and closer in spirit to many modern approaches (Ewing, 2002).

At the same time, literature about Russian (Soviet) schools of this period remains quite limited, and literature about the teaching of mathematics during these years is altogether sparse. In this respect, one must mention the works of Korolev (1958) and Korolev, Korneichik, and Ravkin (1961) — both of which, although far from impartial, make extensive use of archival materials — as well as the small volume by Andronov (1967). Among English-language publications, it will suffice to mention the works of Holmes (1991) and Ewing (2002), which examine developments that occurred in Soviet education during the period in question without examining what took place in specific subjects.

In this chapter, we will be inevitably forced to go beyond the bounds of mathematics education in the strict sense of the word, since it too was subject to new general rules, and consequently, much in it cannot be understood without taking a broader view. At the same time, we are interested in the specific nature of this subject: for all the features that it shared with education in general, mathematics education still existed under conditions that were completely different from those of, say, the teaching of history (Karp, 2007a), and the history of its development is by no means identical to the general history of education in the Soviet Union. In pursuing our investigation, we will aim as much as possible to rely on primary sources, especially archival materials.

---

[1] This and subsequent translations from Russian are by the author.

2 Rejecting Drills and Rote Memorization

Ivan Tolstoy (1858–1916), the minister of education in the administration of Sergey Witte, which followed the revolution of 1905, wrote about the schools of his time:

> Everyone, it seems, agrees that our secondary schools are bad. But when questioned about their shortcomings, when asked what's wrong with them, everyone will give a different opinion (Tolstoy, 2002, p. 330).

According to Tolstoy, parents and students would like to see less of everything and would like everything to be easier. University professors, on the other hand, would complain about the ignorance and unpreparedness of gymnasium graduates. Tolstoy himself cited "encumbering schools with political functions that are harmful to them" (Tolstoy, 2002, p. 334) as the most prominent of the existing shortcomings, but he did not fail to mention others as well, such as overloaded courses and "excessively casuistic problems, calculated to trip up the students or to develop their cleverness, but not very useful for learning mathematics." It should be kept in mind that, despite their proliferation in the 20th century, secondary educational institutions — whether private or government-run — were accessible only to a small percentage of children at that time (Andronov (1967) writes that approximately 5% of children attended gymnasia).

Consequently, after the October Revolution of 1917, which brought Lenin and the Communist Party to power, educational reformers were united in their hatred for the old educational system, with its schools of "drills and rote memorization." However, views about what should be constructed to replace the old system were by no means always identical even among the members of the State Academic Council, a methodological center formed in the early 1920s, that was headed by Lenin's wife, Nadezhda Krupskaya (Blonsky, 1971). Such differences of opinion were still permitted at that time, as were differences between what took place in different parts of the country. Moreover, it should be borne in mind that Russia, which had already lived through a lengthy world war, very soon became submerged in a civil war, which officially ended only in 1920 and which left many if not all parts of the country in utter ruin.

A surviving page from the minutes of a staff meeting at Petrograd's unified labor school no. 15 (School #15, 1921, p. 100) conveys the general atmosphere of many schools at that time. The page's margins are covered with drawings that are obviously the secretary's doodles, while in the middle of the page the following inscription appears: "Causes" (of difficulties, evidently); and below: "1. The cold — they don't come to class. 2. Teachers lack provisions. 3. Shortage of teachers..." and so on.

The recommendations and even the direct orders issued by the ministry of education (the People's Commissariat of Education) were not always carried out, and therefore they cannot be used as a basis for forming a realistic picture of the actual situation in the schools. For example, this is how a school in the center of Moscow is described in a brief article entitled "Around Moscow's Schools" (Lomakin, 1919, p. 17):

> The classrooms are in a state of chaos... Icons hang in some classrooms and the cafeteria. No emphasis is made on labor in theory or in practice. The curriculum remains broken up into subjects.

Such was probably the state of affairs at other schools as well. The official line, however, was that schools should not have icons and that schools should not have subjects. The "unified labor school," which replaced all of the different types of schools that had existed previously, was described in a resolution of 1918 (Abakumov *et al.*, 1974, p. 133). The resolution stated: "Productive labor must be the basis of school life.... It must be tightly, organically linked with an education that sheds the light of knowledge on all of surrounding life" (Abakumov *et al.*, 1974, p. 135). The new communist education authorities naturally rejected all forms of penalizing that had been employed in pre-Revolutionary schools, just as they rejected exams and mandatory homework assignments. The division of the curriculum into separate subjects also had no place within the new conception of education.

Genuine hatred for the oppressive routines of czarist schools, reflected in the memoirs of the most prominent educators of the time (Blonsky, 1971; Shatsky, 1929) — along with the desire to create a "new, free man" — spurred a search for fundamentally new pedagogical methods and approaches. The principles and methods of progressive education — well-developed by that time in the United States, with its child-centrism, the notion of learning by doing, and projects — turned out to be consonant with Soviet education's new objectives. Dewey, Kilpatrick, Counts, and other leaders of American progressive education visited the USSR during the 10–12 years following the Revolution and expressed admiration (even if it was not entirely without reservations) about what was taking place in Soviet education (Ravitch, 2000). Their works, in turn, were indispensable sources for Soviet reformers (even if their admiration was also not entirely without reservations). The role and position of mathematics under these new developments turned out to be quite different from what they had been previously.

3 The Place of Mathematics in the New Schools and the Objectives of Mathematics Education

"The course in mathematics — in terms of its basic minimum program — is constructed and conducted not so much in the interest of future mathematicians or future engineers, chemists, statisticians, and so on, as with the aim of filling in those missing links within the system of liberal education which only mathematics can provide": thus stated the model program of the Northern Region (Komissariat narodnogo prosvescheniya, 1919, p. 5). O. A. Volberg, the head of the natural sciences and mathematics division of the Department of School Reform within the People's Commissariat of Education, wrote: "The labor method places mathematics education in close connection with the productive capacities of the workers' commune" (Volberg, 1918a, p. 14). The program of 1920 explained that the main significance of mathematics resided in the fact that it "provides methods that are irreplaceable in their application to reality," while the "value of mathematics as an abstract science — as a form of mental gymnastics for the young — is worthless" (cited in Korolev, 1958, p. 258).

Meanwhile, in Moscow, Chistiakov (1918), who collaborated closely with the Department of School Reform (Andronov, 1967), examined the two aims that mathematics education usually sets for itself — the "formal mental development of the students" and the "material significance of mathematics education, i.e. the value of the direct application of mathematical knowledge to practical life" — and reached the conclusion that neither of these objectives is achieved in practice, and that therefore neither of them "can be established as a direct goal for mathematics education" (Chistiakov, 1918, p. 6). The purpose of mathematics education, according to Chistiakov, must be to "inspire students to think mathematically": it must "instill interest and love for mathematics in the students and promote independent work in the study of mathematical truths" (Chistiakov, 1918, p. 6).

According to one approach that emerged at the time, mathematics was nothing more than a useful method for solving various real-world problems. Another approach saw it as a separate subject, while nevertheless arguing that its teaching should be reformed. In an article entitled "Two Worldviews", O. A. Volberg (1919) polemicized against the latter point of view. After the members of a Moscow mathematics circle criticized plans for a model mathematics curriculum proposed by his commission, he described them as individuals who were "locked within the bounds of self-sufficient scientific thinking, constructing abstract schemas in the solitude of their offices" (Volberg, 1919, p. 84). According to Volberg, mathematical abstraction did not precede practical application, but on the contrary emerged from it in a natural fashion. Consequently, in response to objections that the new schools would require teachers with an encyclopedic breadth of knowledge, since only such individuals would be capable of introducing new ideas "from various realms of application into the students' curriculum," Volberg argued that, on the contrary, teachers would merely have to get involved in "practical work in various spheres of the construction of life."

> And who will teach mathematics, who will teach physics, who will teach botany? No one. Recall that we have agreed that the teaching of subjects should have no place in the schools. Every teacher–worker will apply mathematics, physics, biology, etc., at such times and to

> such an extent, in such contexts and to such an extent, where this is required for effective teaching. And only then (Volberg, 1919, pp. 84–85).

In the article cited earlier, however, Volberg (1918a) did state that "the idea of functional interdependence is the linchpin that must give unity and stability to mathematics," that "it is necessary to accustom children to equations from the very first year of schooling," that "it is necessary to underscore and insist on the connection between the origins of mathematics and the history of its development" (Volberg, 1918a, p. 15), that students in upper classes must "become acquainted with the foundations of calculus and continue to study analytic geometry in close connection with the natural sciences" (p. 16). How this whole program, which was elaborated by the international reform movement, was to be realized by a teacher whose only preparation lay in "sanitation, agricultural work, and various other branches of the productive and processing industry," as the author carefully specified (Volberg, 1919, p. 84), is something that, unfortunately, was left unexplained.

4 School Programs and School Practice

The same Volberg (1918b, p. 35) also commented ironically that, after humoring themselves with talk of new schools based on "independent work, creativity, and labor," teachers are forced to go back to using the old, pre-Revolutionary textbooks in their teaching. The 1918 resolution on the labor school promised that new model curricula would be published to help educators in their work. It also conceded that teaching plans had to be quite flexible in their application to local conditions and even permitted the "introduction of various subjects for particular groups of students... on the condition that the unity of the course be preserved" (Abakumov *et al.*, 1974, p. 135). New programs gradually began to appear, but it should not be forgotten that they were applied in a "flexible" fashion.

The 1918 plan for a model curriculum in mathematics (Proekt, 1918) was quite intensive. For example, the fifth grade program included (but was not limited to):

> Formulating linear equations with two variables; solving them graphically and analytically. Graphically representing the functions $y = ax^2$ and $y = ax^3$ ... Square and cube roots. Finding square roots by division and by trial and error.... Graphically representing the function $y = a\sqrt{x}$.... Sine and tangent as functions of the acute angle. Solving right triangles using sine and tangent tables. Deriving the formula $\frac{a}{\sin A} = \frac{b}{\sin B} = \frac{c}{\sin C}$ (Proekt, 1918, p. 42).

Programs that appeared later were more modest. Even so, for example, the State Academic Council's curricula for the ninth grade, printed in Krogius and Popov's (1928, p. 55) manual, include progressions; concepts involving limits; the volumes of prisms, pyramids, cylinders, cones, and the sphere; the elements of trigonometry; exponential and logarithmic equations; Newton's binomial theorem. The course devoted a large amount of attention to functional interdependence; consequently, it recommended applying the conception of "motion" in geometry; called for early, propaedeutic instruction in geometry; and actively encouraged a "genetic" form of exposition (today we would use the word "inductive"), in which concepts and propositions are presented gradually, as progressive generalizations.

Krogius and Popov (1928) provide an extensive bibliography, but do not single out any recommended textbook. They do make methodological recommendations, but in doing so they also often mention the existence of various alternative points of view. For example, they write that the "explanatory note" to the State Academic Council's programs "does not consider it feasible to provide for a more detailed study of functions by introducing the concept of the derivative.... However, many methodologists ... believe that it is necessary to introduce these concepts" (Krogius and Popov, 1928, p. 95). And they list the textbooks that follow such a program.

The wish to study mathematics merely as a useful instrument in a labor-related context, which led educators to avoid isolating mathematics into a separate subject, was reflected in a special kind of propaganda, or, to put it more accurately, in the propagation of a complex system of teaching. Within this system, students did not study mathematics or, say, botany, but explored some general theme or real-world situation: "harvesting crops," "sources of energy," "transportation," and so on, or even more modest topics, such as "the duck" (Kaverin, 1950). The topic was studied from the perspective of various school subjects, including mathematics. For example, Ponomareva (1926) offers a large set of various different assignments on the topic of "The Post Office" for grades 3–5. These include problems on measuring and determining distances, working with tables, reading and constructing diagrams, calculation problems, and word problems. The methodology textbook edited by Leifert (1931) was not restricted to such modest topics, but showed how to prepare a class on "getting ready for International Red Aid day." After formulating the overall aim of the lesson — "to educate the children in the spirit of international class solidarity" (Leifert, 1931, p. 71) and so on — the textbook offered a great many assignments, from which the children could learn about the sorry lot of day-laborers in Spain; compute how much even the best-paid worker in Turin earns, in rubles; construct a diagram showing the relation between the salary of a British worker and a worker in the British colonies, etc. The problems were classified by topic: "salary," "unemployment," "the strike movement," "White terror," "the growth of the Communist Party's influence," "International Red Aid," "USSR nationalities policy"; and students were expected to work on them collectively, in seven teams (Leifert, 1931, pp. 72–86).

The State Academic Council's programs specifically stipulated that the acquisition of mathematical skills during the first 4–5 years of schooling must take place in the context of working on real-world materials and not in special classes. As for the upper grades, mathematical knowledge was to be acquired in the process of studying physics, mechanics, and astronomy (Gratsianskii, 1924, p. 151). Even here, however, certain allowances were made. In beginning classes, specialized exercises could be given in addition to the complex lessons once students had appreciated the need for such specialized exercises while studying complex topics. In the upper grades (high school), the programs went even further: "the complex system here consists in the fact that, in lessons devoted to various subjects, the same question is simultaneously discussed from many different points of view" (Gratsianskii, 1924, p. 152). Thus, the need for a separate school subject — at least in classes above the beginning level — was, however reluctantly, recognized.

Gratsianskii (1926) observed that "teaching based on life experience" sometimes led to having students solve meaningless problems based on real-world facts. As an example, he offered the following problem, which students were given while studying the topic "the horse": "One horse needs four horseshoes. How many horseshoes do 23 horses need?" (Gratsianskii, 1926, p. 9). Such problems, in Gratsianskii's opinion, contributed nothing either to the study of the overall topic itself or to the development of students' computational skills. Gratsianskii went on: "The State Academic Council programs' position that 'mathematics is studied in the process of' studying other topics must be understood as a proposal to conduct mathematics classes 'in any which way'" (Gratsianskii, 1926, p. 9).

Korolev, Korneichik, and Ravkin (1961), however, argue that in the actual practice of teaching mathematics during this time, particularly in the upper classes, often no special attempts were made to follow the general recommendations, not even in their milder versions. For example, algebra and geometry were taught not within the framework of a singe subject, but separately; nor was it always necessary to use any available means to connect separate disciplines into a single complex subject (Korolev *et al.*, p. 176).

5 Discussion: Some Conclusions

It is not our purpose to praise or criticize what went on in the schools during the period described. It is evident that providing students with a deep and durable knowledge of mathematics was not seen as a top priority for the schools at this time. The creators of the existing educational system saw its virtues elsewhere. At the same time, it should be recalled that it was this educational system that produced the creators of the Sputnik — rather than the educational system embraced in the 1950s, which educators in other countries rushed to study and imitate after the satellite went into space (Karp, 2009a). Kolyagin (2001), who ridicules the post-Revolutionary utopian dreamers, nonetheless prodigiously praises I. K. Andronov (he "molded me into a scientist," Kolyagin (2001, p. 179) writes). Meanwhile, Andronov (1967, pp. 99–108) himself described how he worked under the leadership of O. A. Volberg, for whom he expresses great respect. Many methodological ideas that were popular at the time (for example, the attempt to organize research in class or to present materials in a "genetic" fashion, so that students would go through a "natural sequence of concepts" (Gratsianskii, 1926, p. 11) instead of obtaining knowledge in its final form) exerted an influence on later teaching.

A "flexible" attitude toward instructions and recommendations, however, also meant that new ideas did not always penetrate into the schools in accordance with educators' plans. The "Report on mathematics education in Leningrad schools for 1923–1924" (Otchet, 1926) states that although students everywhere had indeed begun to study geometry during the first five years of schooling, in accordance with official recommendations, only 67% of the schools began exposing students to its elements in the first grade ("grade A," as it was called at the time), while, say, 16% did this only during the third year of schooling. Already at the close of the period discussed, the prominent educator Pistrak (1929) wrote that "the old objectives of mathematics in the schools were not achieved, are not achieved, and cannot be achieved. The new objectives have thus far been formulated only in the official programs and have not yet been implemented in practice" (p. 5).

It should also not be forgotten, that widespread academic failure was a common phenomenon. Blonsky wrote in 1928 that an inspection of a number of Moscow schools during the preceding year revealed that, on average, during the "first stage" (roughly, elementary school), 27% of children failed two or more subjects (usually writing and arithmetic). Furthermore, in some "second stage" groups, all but two or three children failed at least one subject (Blonsky, 2006). Not the least important reason for these failures was likely the teachers' lack of preparedness. Voronets (1928, p. 3), the author of a mathematics methodology manual, wrote that teachers' courses and conferences focused mainly on fundamental questions of school reform, "while at the same time questions about subject methodologies were pushed aside." Once again, it should not be forgotten that teachers' teachers included individuals like the above-cited Leifert, a person who was absolutely mathematically illiterate (Venttsel and Epstein, 2007), but quite ideologically active.

We cannot analyze the different versions of the recommended programs systematically here, although they did undoubtedly change. Korolev, Korneichik, and Ravkin (1961) believe that the teaching of mathematics steadily improved during the 1920s. The age of experiments, reforms, liberties, and exploration, however, was coming to an end.

6 The CPSU's Resolutions on Education

Between 1931 and 1936, the Central Committee of the Communist Party issued a number of resolutions — *On elementary and secondary schools* (1931), *On the educational programs and regulation of elementary and secondary schools* (1932), *On the structure of elementary and secondary schools in the USSR* (1934), *On the organization of teaching and internal regulation in elementary, middle, and high schools* (1935), *On pedagogical perversions in the system of the People's Commissariats of Education* (1936), and others — which fundamentally transformed the Soviet education system. Noting that the number of students in elementary and secondary schools had risen from 7,800,000 in 1914 to 20 million in 1931 (Abakumov *et al.*, 1974, p. 156), the 1931 resolution remarked that the schools' principal shortcoming consisted in the fact that "teaching in the schools does not provide students with a sufficient breadth of general knowledge and does not satisfactorily solve the problem of preparing for vocational schools and colleges sufficiently competent individuals with a sound grasp of the fundamentals of science" (Abakumov *et al.*, 1974, p. 157). Consequently, the resolution called for an increase in the amount of knowledge being transmitted to students, a more systematic approach to its exposition, greater methodological guidance for teachers, and more generally the establishment of a rigid organizational structure that regulated both the work of the school and the entire educational process. For example, the resolution stipulated that "in every separate subject there must a single, mandatory textbook, approved by the

People's Commissariat of Education of the RSFSR and published by the State Pedagogical Publishing House" (Abakumov, 1974, p. 165). The work of psychologists, special educators, and all so-called "pedologists" (pedagogical psychologists) was declared to be a harmful and anti-scientific perversion, which sought to characterize normal and even gifted students as "difficult." The complex system or projects method was definitively rejected: henceforward, this "eructation of the left" was to be unconditionally eradicated.

The resolutions regulated virtually every detail of school life. Volin, the head of the Central Committee's Department of Schools, spoke enthusiastically about the party's attention to trifles at an educators' conference in Leningrad in 1935:

> Comrade Stalin takes an interest in and occupies himself with all of these things, signs resolutions that specify the form and size of the pens that should be used, how sharp the end — the tip — of the pen, should be (GK VKP(b), 1935, p. 55).

A great deal of attention was also devoted to regulating school life — to organizational and ideological details — at more local levels. Memos about the implementation of the Central Committee's resolutions (OK VKP(b), 1936a, b) were filled with reports that not enough children had yet been transferred from special education schools to ordinary ones, that pedological literature had not been eliminated everywhere, that somewhere trash had been found on the floor, that the duration of a class had been altered (or had not been altered) in the appropriate fashion, that not enough efforts were being made to eliminate dirt from students' notebooks, and so on. But significant changes were also implemented in the substantive methodological–mathematical sphere.

7 Changes in the Programs and Organization of Mathematics Education

The earlier flexibility disappeared rather quickly. An instructional letter issued by the People's Commissariat of Education in 1933 (Berezanskaya, 1933) contained a recommended schedule of classes in different mathematical subjects; and although it conceded that the schedule was only approximate and "must be more precisely determined by teachers themselves in accordance with the concrete conditions of their work" (Berezanskaya, 1933, p. 29), it was accompanied by critical accounts of those who had rearranged the order of the subjects or had deviated from the program by telling the students more than they were supposed to (p. 67), and thus could not help but produce the impression that classes had to be scheduled exactly in the way that the instructions specified.

The actual contents of the programs were also altered. Berezanskaya (1933) emphasized that it was improper to talk about functions in grades 6 and 7. Beginning with the 1934–1935 school year, the elements of analytic geometry and the analysis of infinitesimals were eliminated from the program, as was the notion that the idea of functional interdependence must be cultivated throughout the entire course. The history of mathematics practically disappeared from the curriculum (Scherbina, 1938). Analyzing the programs of the 1937–1938 school year, Sakharov (1938, p. 78) wrote: "With a single stroke of the pen, the propaedeutic course in geometry for the fifth grade has been eliminated — a course for which more than one generation of mathematicians had fought."

These changes were motivated by the fact that students were overloaded and therefore failed to assimilate the basic points of the course (in the Central Committee resolution of 1932, the propaedeutic study of stereometry in seventh grade was mentioned as an example of the fact that "a number of subjects are covered hastily, and the children fail to acquire a sound grasp of the relevant knowledge and skills" (Abakumov *et al.*, 1974, p. 161)). What is noteworthy, however, is that the programs were thus brought back to their pre-Revolutionary form, and indeed to the form they had prior to 1907, when reformist tendencies began to penetrate into Russian education (Scherbina, 1938). Moreover, slightly revised versions of pre-Revolutionary textbooks were brought back as standard textbooks to be used in all schools (the most important of these was Kiselev's text, Karp, 2002). Exams, which had previously been categorically condemned, were also revived (Karp, 2007b).

These changes were not implemented overnight. For example, as late as 1938, the use of Kiselev's textbooks as the unique, standard textbooks in algebra and geometry was still considered only a temporary measure — it was expected that a variety of new textbooks would be developed gradually in every subject, tested out in practice, subjected to wide discussion, and that only then would standard textbooks be chosen among them (Stabil'nye uchebniki, 1938). The time for such discussions, however, was already coming to an end. The introduction of exams also went through several stages before a system of absolute centralization took shape — already after the war — in which problems for graduation exams across the country were written in Moscow. Every year brought certain new changes; it is not possible to analyze the complex sequence of their implementation here.

It should be noted, however, that even a decade after the mathematics curriculum had been restructured, not everyone considered the changes beneficial. A report from the Leningrad City Institute of Teachers' Continuing Education (IUU, 1945) gives a list of the principal existing shortcomings, such as the fact that

> even in the minds of the teachers, the courses [in algebra and trigonometry] consist of disconnected scraps, while students almost never grasp these subjects as unified wholes. The nature of the programs and textbooks contributes to this effect. Thus, for example, the idea of functional interdependence, which should penetrate the entire mathematics curriculum and unify all of its sections into an orderly system, is taught to the students in bits and pieces and only at the end of the eighth year is 10 hours allotted to it, as if in order to systematize the knowledge that has been acquired.

Among other measures whose time has come, the author of the report names the introduction of the elements of analysis into the curriculum, and, most importantly, the elimination of obsolete sections from the program (such as Newton's binomial, or the secant and cosecant in trigonometry). He also complains about the absence of a propaedeutic course in geometry for grades 5 and 6. Finally, in conclusion, he expresses his views on textbooks:

> One of the principal causes of the shortcomings in mathematics education in its current state resides in the failure of existing textbooks to meet even minimally reasonable requirements. These textbooks

> and problem books were written 50–100 years ago, and the revisions that they have gone through have only made them worse. It is necessary to have not merely one standard set of texts, but several parallel authorized sets, so that schools might be free to choose among them. It is necessary to encourage creativity in this direction, which is not being done at the present time (IUU, 1945, p. 16).

In this way, the report argues (futilely, at that time) for a return to certain reformist methodological ideas that had been popular during the 1920s.

Apart from the rejection of these ideas, the 1930s witnessed many other changes in education programs. For example, ninth grade curricula for 1935 contain topics such as progressions, the generalization of the concept of the exponent, exponential functions and logarithms, inscribed and circumscribed polygons, the concept of the limit, the circumference and area of a circle, the relative positions of straight lines and planes in space, and the elements of trigonometry. The difference between these curricula and the State Academic Council's programs described above is evident. In 1935, the ninth grade was no longer the last grade of school: there was now a tenth grade, where students covered topics such as combinations and Newton's binomial theorem or volumes. The programs in algebra, geometry, and trigonometry were given separately in 1935 — these were now three separate subjects, and not three components of a single subject ("Mathematics").

But, to repeat, what was even more important was the fact that methodological guidelines became far more rigid, and they had to be followed to the letter — over the years, the requirements became more and more strict. In 1952, participants at a methodological conference (IUU, 1952, p. 21) spoke approvingly about how the tone of the school program had changed in accordance with one of the guidelines: initially, following the program was optional; later, following the program became mandatory; and lastly, the program was simply followed.

The behavior of the teacher in the classroom — the precise manner in which the established program was supposed to be taught in practice — was also centrally regulated:

> Homework must be checked during every class for 10–15 minutes. . . . The teacher must call the student up to the blackboard, take his notebook, and look through it quickly, pointing out mistakes to the student if they are minor. If the teacher sees that the student needs additional instruction, he should arrange a "working with failing students" session (Berezanskaya, 1933, pp. 11–12).

In this way, methodology assumed an increasingly normative aspect.

8 Rising Demands and Growing Numbers of Students

As stated above, in many respects school programs and school practice returned to pre-Revolutionary models. But what had earlier been offered only to a relatively small fraction of the population now became accessible to an incomparably greater number of students. A speech by the head of the Central Committee's Department of Schools, Volin, conveys an idea of the growing numbers of students who were graduating from secondary schools:

> This year, 40,000 children graduated from ten-year schools, and a significant number of them went on to matriculate at colleges. In 1936, 70,000 children will graduate from school, and in 1937, 120,000 children will graduate from secondary school — almost as many as there are places in colleges (GK VKP(b), 1935, p. 54).

A Leningrad city school board report (LenGorONO, 1938a) for the 1937–1938 school year gives the following statistics about the number of students in the city's schools, by year and grade:

	Grade			
School year	7	8	9	10
1935–1936	22,997	10,993	6,211	2,500
1936–1937	26,984	14,328	7,342	4,533
1937–1938	34,074	19,411	11,360	6,461

If earlier a rather sizable proportion of the student body was considered unfit for a serious curriculum, then now former techniques for determining intellectual ability through tests and questionnaires were categorically rejected. Zhdanov, party ideologue and secretary of the party's Leningrad Regional Committee, explained this to educators in the following way:

> This system of tests has been rejected by the Central Committee, since as you recall, in its decision concerning schools the Central Committee long ago proposed eliminating all systems and forms of keeping track of students' progress that are harmful and unnecessary (GK VKP(b), 1936a, p. 9).

In this way, any kind of preliminary selection of students for a serious and challenging course in mathematics was ruled out. Meanwhile, the requirements in mathematics continued to grow. The growing requirements and achieved success rates were pointed out constantly:

> Relatively recently people constantly said things like: "you know, in sixth grade, it is hard to do proofs." If a factorization of $a^2 - b^2 - 2bc - c^2$ was called for, then two years ago this seemed very difficult. Schools avoided such assignments. Today, there is no school in which the students would be unable to factor this expression (LenGorONO, 1938b, p. 5).

And the requirements continued to become more and more demanding. For example, a methodologist from the Leningrad City Institute of Teachers' Continuing Education compared the situations in 1940 and 1949 in a tone that was positively exultant:

> In 1940, tenth grade exams in algebra contained problems that required students to formulate a quadratic equation with integer coefficients and the students were not asked to provide a detailed explanation and derivation of the equation, such as is now given not only by tenth-graders, but by seventh-graders as well. In 1940, students were not asked to explore the solutions to algebraic word problems; now they are, and in a rather serious fashion. Problems in geometry were solved without justification and in this way the depth of the students' grasp of theory was not tested; now, the explanation of a problem's solution determines the quality of the solution; and

> if in 1940 students could be given an A for a correct solution without an explanation, then now they would get a C for such a solution. If we say that students are given too few construction problems and problems involving proofs, then previously they were given no such problems at all. In recent years, attention to the theoretical issues in the course has increased, and as a result changes are being made more consciously in all sections of the course. Oral exercises, which were unheard-of before, have become standard practice in all sections of mathematics for a large number of teachers (IUU, 1949, p. 26).

Shortcomings in student preparation that were identified at the same time are discussed below. In general, it is extremely difficult to compare the results achieved during different periods objectively. But the growth in the requirements in mathematics after the Central Committee's resolutions is unquestionable. To give just one example, here are two versions of the Moscow State University entrance exam, the first from 1928 and the second from 1953:

Moscow State Univesity entrance exam from 1928

1. Find the volume of a regular, four-faced pyramid, given a dihedral angle α at its vertex and the area of a face.
2. Factor $x^8 - 4y^8$.
3. Write down six numbers between 7 and 35 that form an arithmetic progression with those two numbers (Gurvits, Znamenskii, and Fridenberg, 1929, p. 78).

Moscow State Univesity entrance exam from 1953

1. Two factories received orders for the same number of cars. The first factory started operating 20 days earlier than the second one, and finished 5 days earlier. By the time that both of the factories together had completed one third of the combined total of their orders, the first factory had manufactured 4 times as many cars as the second factory. All together, the first factory operated x days, manufacturing m cars per day, while the second factory operated y days, manufacturing n cars per day. Find all values of x, y, m, n and

all relations $\frac{x}{y}$ and $\frac{m}{n}$ that can be determined from the conditions of the problem.

2. What values of x satisfy the inequality $\frac{1}{x} < \frac{1}{x-1} - \frac{1}{2}$?
3. How many six-digit numbers are there that have three even digits and three odd digits? (Modenov, 1954, p. 77).

9 The Struggle to Increase Teachers' Mathematical Knowledge and the Role of Research Mathematicians

It is notable that in tackling the problem of preparing students for colleges (that is, for working in heavy industry — and first and foremost, in the defense industry — that was being established in the Soviet Union), the leadership of the country considered it necessary to raise the level of teachers' mathematical preparedness significantly. Aleksinsky, who oversaw education in Leningrad, expressed himself as follows:

> It must be said bluntly: in our education system, not yet everyone has real knowledge of their subject.... I will say why our people do not have knowledge: some of them graduated from college before the Revolution and over the intervening years they have not kept abreast — could not for many reasons keep abreast — of science, and they have fallen behind science. Many have even graduated from the university and other educational institutions — teachers of physics and mathematics — but everywhere they have simply fallen behind modern science. They must catch up. They must simply be honest and say: I can't step aside, I can't fall by the wayside — they must catch up. Others studied during the years of the Revolution, when in the mathematics department they were taught to help the factory to increase the productivity of labor, in the history department they were taught to help restructure collective farms, and so on, and in vocational schools they were not taught grammar, but occupied themselves with community service cleaning vacant lots, and so on. This must be admitted — there's nothing else to do.... There's nothing we can do. After all, it is only in recent years that education has started to acquire some kind of coherence — and still not much — in our colleges and vocational schools (LenGorONO, 1936a, p. 85).

At another conference, Leningrad city school board representative Popov told mathematics teacher educators:

> Face the facts. What do you teach the teacher? You yourselves have stated that you teach him methodological techniques. But is this enough? No, it is not enough. Since the teacher does not know his subject, he must acquire command of his subject. Otherwise, his teaching will be mere scholasticism, mere going-through-the-motions, which the Soviet government has no need for (LenGorONO, 1934, p. 28).

Consequently, professional mathematicians became more and more actively involved in working with the schools. Certain traditions of doing educational work had long existed among the intelligentsia, and these traditions undoubtedly helped to sustain an interest in working with schoolchildren among a number of mathematicians (for example, Andrey Kolmogorov — who would become one of the greatest Russian mathematicians of the century — had worked at a secondary school as early as the very beginning of the 1920s). However, during this new phase the involvement of research mathematicians rose to an entirely new level. "Together with many Moscow scientists, I was invited to attend the graduation examinations at one of the schools," wrote the academician Luzin in his article, "Pleasant Disappointment," published in June 1936 in one of the most important newspapers in the country, *Izvestiya* (Demidov and Levshin, 1999, p. 253). Subsequently, the praise that Luzin lavished upon the school and how mathematics was taught in it formed the pretext for the beginning of a campaign against him — one should not praise, he was told, but uncover shortcomings, thus helping the school to improve. But what was typical of the period was the fact that an invitation to attend graduation exams had been extended to a person who was so far removed from the world of the schools.

The transcripts of a meeting between People's Commissar of Education Bubnov and Leningrad research mathematicians (LenGorONO, 1936b) reveal certain communication problems between the authorities and the research mathematicians. In response to criticisms of the textbooks, Bubnov repeatedly and insistently asks the scientists to point to any crude mistakes in them: "Give an example, take some theorem, say that it is formulated incorrectly" (LenGorONO, 1936b, p. 61). The mathematicians, by contrast, see their role as consisting not merely in tracking down mistakes, and they attempt to explain that the presence or absence of mistakes is not the whole story. (Prof. Tartakovsky ironically comments that Tolstoy's *War and Peace* is a book without mistakes, but one cannot use it to teach mathematics (LenGorONO, 1936b, p. 46)).

Prof. Fikhtengolts, offering a sharp criticism of Gurvits and Gangnus's (1936) geometry textbook, remarks:

> The tragedy of the situation has nothing to do with these mistakes, each of which can be individually fixed. The tragedy resides in the fact that the textbook is written in a way that can ruin the students instead of teaching them, instead of developing their logical reasoning. So I repeat, one can give as many examples [of such mistakes] as one likes, but that's not the point. The state of things is much worse than can be imagined... the teacher cannot work with [this textbook], and when he begins to present things differently, then he starts getting reprimanded — you don't have the right to deviate from the textbook (LenGorONO, 1936b, p. 62).

During such discussions, pre-Revolutionary textbooks were usually offered as examples of the correct way of doing things, and it was to these textbooks that educators gradually returned. The research mathematicians criticized the preponderance of mindless routines in the schools, the absence of difficult problems, the fact that difficult and substantive sections had been removed from the textbooks. All of these discussions had a definite influence on teaching practice in the schools. In this connection, one must mention the appearance of mathematics olympiads (in 1934 in Leningrad), with the aim of identifying and attracting gifted students to mathematics, the formation of a system of mathematics clubs ("circles"), and the publication of books and pamphlets for those interested in mathematics (Karp, 2009b).

At the same time, the direct employment of research mathematicians in teacher education often turned out to be not very useful. A report from the Leningrad City Institute of Teachers' Continuing Education for 1938–1939 stated:

> A rather large number of major research mathematicians have a negative attitude not only to the mathematical methodologists who are members of the department, but to the methodology of mathematics itself. At the same time, as a rule the research mathematicians have an extremely poor knowledge of secondary schools, of the conditions under which teachers in them must work, and of teachers themselves (IUU, 1939, p. 3).

As a result, it was not feasible to use research mathematicians in working with teachers directly: "The teachers remained dissatisfied and discontented" (IUU, 1939, p. 3).

10 What a Lesson must Accomplish

And yet, the subject-side of working with teachers predominated. At Zhdanov's conference (GK VKP(b), 1936a), already cited above, existing pedagogical manuals were constantly criticized for being too general and not providing concrete recommendations about specific subjects. It was not merely and not even primarily general pedagogical principles that became the center of teacher education, but precisely subject methodology: the discussion concerned how to teach the concrete subjects that were studied in school.

The lesson was seen as the main form of instruction and consequently a very high level of intensity was demanded of it. The Leningrad city school board's briefs and reports noted that

> The principal defect of many lessons is insufficient mathematical content. The teacher does not know how to organize his work so as not to waste a great deal of time (LenGorONO, 1936c, p. 31).

It was emphasized that rote memorization and routine training were not the aim. "Insufficient attention to deliberately learning the content of the curriculum" was named as another shortcoming. The lesson was criticized for the fact that

> the exercises and problems that are given to students are very simple: there is nothing to think about. Such work cannot attain the main

> objective of mathematics education: the ability of think and reason correctly. Furthermore, it does not teach students to apply theoretical knowledge to solving exercises and problems, and reduces their interest in mathematics (LenGorONO, 1936c, p. 31).

Aleksinsky, the Leningrad city school board chairman cited above, elucidated the problem facing teachers in the following way:

> What do we want when we demand from the teacher a lively, active approach, vividness, examples, and so on, and so forth? We want all of these measures to secure a lively interest on the part of the child in the subject being taught (LenGorONO, 1936a, p. 84).

City school board officials who visited classrooms noted that "the following plan is typical of most lessons: (a) homework review; (b) presentation of new content; (c) content reinforcement; (d) homework for the next lesson." But they immediately added that "the different steps of this plan are carried out in extremely varied ways" (IUU, 1949, p. 13).

For example, an eighth-grade lesson in geometry — presented as a successful model — was described in the following way (LenGorONO, 1936c). The lesson began with one student getting called up to the blackboard in order to solve a computational problem from the homework assignment, and another getting called up to prove a theorem that had been covered earlier. While they were doing their work, very carefully making diagrams and notations, the class discussed the answers to questions such as: "Is it true that all isosceles triangles are similar? That all isosceles right triangles are similar? Are isosceles triangles with a base angle of 35° similar?" and so on. After this, the two students at the blackboard presented their work. And then the class turned to new material — proving a theorem about the relation between the areas of similar polygons. The proof was almost entirely constructed by the students themselves, who responded to the teacher's questions. Aside from posing these questions, the teacher did not help the students in any way. The students were given the theorem that they had analyzed and computational problems as a homework assignment. The observer noted the clarity and precision of the teacher's questions and her concern for the mathematical development of the students, which manifested itself in the selection of substantive problems, the careful construction of the lesson, and its orderly unfolding (LenGorONO, 1936c, pp. 44–45).

By contrast, observers noted these main kinds of shortcomings (the list below draws on a transcript from 1954, but analogous remarks and demands were expressed during the 1930s–1940s):

1. Insufficient attention to the conceptual side of the lesson.
2. Inability to plan and coordinate the separate parts of the lesson in time.
3. Insufficient activation of the knowledge assimilation and reinforcement process.
4. Insufficient attention to teaching the students independent work skills.
5. Inadequate attention to repeating what has been covered and to visual models in teaching.
6. Inability correctly to determine students' knowledge through questioning.
7. Lack of attention to students who are falling behind in class.
8. Gap between theory and practice, poor implementation of the principles of polytechnic education in the teaching process (IUU, 1954, p. 18).

The ideal lesson was seen as being challenging, but capable of teaching even the student who is falling behind; varied in its pedagogical and methodological techniques and aims; implemented through carefully chosen mathematical problems; developing the students' abilities, but also aimed at giving them a firm grasp of the main skills and knowledge prescribed by the program; expecting the students to engage in work that is active and independent, but at the same time clearly and even rigidly steered by the teacher. For the interested and gifted, such a lesson might continue in extracurricular work. Observers' reports often contain special praise for teachers whose students have performed successfully in mathematics Olympiads.[2]

11 The Fight Against Formalism and for the Practical Application of Mathematics

The schools' objective was to have the students consciously assimilate the course in mathematics; consequently, it was recommended that merely formal learning — meaningless rote memorization — be eliminated. Examples of such empty learning were abundant:

> An eighth-grade student from school no. 312, who responded well to questions that she had been specifically prepared for, was asked: "What is the sine of a 30° angle?" and replied: "One half of the hypotenuse." A ninth-grader quickly and correctly stated the definition and described the properties of infinitesimals, and when he was asked to give an example of an infinitesimal quantity, he said "one-millionth," thus revealing his complete lack of understanding of the essence of the question (OK VKP(b), 1947, p. 56).

Another report states that "many students were unable to answer questions that they had not been specifically prepared for or questions posed in an altered form" (LenGorONO, 1946, p. 4), and once again it provides examples — one student was unable to demonstrate the actual relation between a meter and a tenth of a meter, using a concrete example, and so on.

Questions such as the ones cited above, to some extent, themselves served as a means of fighting against merely formal learning. The Ministry of Education's decree (MP RSFSR, 1948) underscored the

[2] Ewing (2002), describing a teacher who considered a lesson successful if at the end of it students would ask what else they could read about the topic in order to enrich their knowledge, concludes that this demonstrates that teaching was aimed only at learning a "clearly defined body of knowledge," and not on developing reasoning ability, since otherwise the students would have been asking for something that might "challenge what was taught in class" (p. 182). At least with respect to the mathematics lesson, such a conclusion appears unconvincing.

role of problem solving:

> It is likewise recommended to have students solve mathematical problems not merely of a low or medium level of difficulty, but of a high level of difficulty as well, and all efforts should be made in the process to develop the students' mental agility, their ability to solve the problem in the most economical, rational manner (MP RSFSR, 1948, p. 8).

The same objective was aimed at by the demand that students provide explanations in their solutions to problems — oral problems in class work and written problems on tests (this, however, led to the development of a new — and far from rational — tradition of making the explanations as long as possible; the Leningrad methodologist Depman, for example, would relate how the solution to a geometry problem in an exam took up 19 pages (Matematika, 1948)).

The ability to apply mathematical knowledge while solving practical problems was usually examined not in connection with the fight against formalism, but as a component of so-called polytechnic education. This education, which had only recently been considered of paramount importance, was practically eliminated following the Central Committee's resolutions. The Leningrad city school board report of 1932 (LenGorONO, 1932, p. 3) stated explicitly that "polytechnization of the curriculum is not sufficient. Many teachers have interpreted the problem of subordinating labor to educational objectives as one that may be solved through the elimination of polytechnicism." During the second half of the 1940s, the idea of "polytechnization" once again became popular. Educators began to argue that a major shortcoming in the instruction offered by schools lay in the abstractness of students' knowledge, in their disengagement from concrete practices — from the productive activity of human beings — and in the empty verbalism in their studies (IUU, 1952, p. 8).

Consequently, somewhat more attention began to be devoted in the school curriculum to so-called practical work — mainly, exercises involving measurement. Interestingly, in the course of discussions about how to make the course in mathematics more "polytechnic," ideas about the importance of studying functional interdependencies — which had been popular in the 1920s — arose once again (IUU, 1953). It appears, however, that the ideas of "polytechnization" began to exert a significant influence on the organization of mathematics education in the country only later, at the very end of the 1950s and the beginning of the 1960s.

12 Student Failure and the Struggle Against It

Despite all efforts, failure was widespread in the schools and large numbers of students were held back. At a party city committee conference at the beginning of the 1936–1937 school year (GK VKP(b), 1936b), it was noted that over the previous year in Leningrad a total of 50,000 students in all grades had been held back. A large share of these failing students came from mathematics classes. A 1936 report (LenGorONO, 1936c) stated that failing students in mathematics over the first quarter in middle and upper grades constituted from 7% (in a class on trigonometry for tenth-graders) to 17% (in grades 5 and 6) of the class. An analogous situation was observed in subsequent years. For example, in the relatively successful Leningrad school no. 105, the success rate in mathematics in 1954 was 87.4% — 137 students had failed the course (School #105, 1954).

Every school and every teacher were expected to struggle untiringly to raise student achievement rates. Zhdanov, the secretary of the party's Leningrad Regional Committee, formulated the issue as follows: "If, for example, a student has made a mistake, then are we allowed to say that he alone is to blame? Or is the teacher also to blame, as the person who taught this student?" (GK VKP(b), 1936a, p. 21). Zhdanov's listeners immediately acknowledged that the teacher undoubtedly was to blame. Consequently, every teacher and every school had to analyze the causes of low student achievement rates and to develop a plan for fighting against them. In the aforementioned school no. 105, the causes that were thus identified included the fact that work with failing students had not been organized in a timely manner, that monitoring of students' knowledge had been inadequate and poorly structured (and that homework assignments had not been adequately checked), that certain teachers were insufficiently prepared, and that discipline was poor. On the other hand, educators noted the fact that the curriculum was overloaded and that the student makeup of the classes was not always optimal. A number of decisions were made as a result of this discussion: to control and monitor the mathematics lessons and give extra assignments in classes with low and poor achievement levels, to re-check written assignments in such classes, to add an extra hour of lessons in classes with low achievement rates, and to organize Young Pioneer and Komsomol activities to fight for higher achievement rates.

Conducting additional lessons with failing students and constantly monitoring their activity were considered part of the teacher's duties — those who devoted insufficient attention to this part of their work were severely criticized. Work through Young Pioneer and Komsomol organizations also was substantial.

The vice principal of a Leningrad school related that, having determined that approximately 50% of the children in his school had D's, he decided to give all of them the following letter to pass on to their parents:

> Dear Comrade, the teachers' and students' collective of such-and-such a school have undertaken to solve an important problem: to welcome the thirtieth anniversary of the October Revolution with a hundred-percent pass rate. A student at the school, your son so-and-so, member of such-and-such a class, has the following grades in the following subjects. We ask and urge you to take measures to make your son work systematically and finish the quarter with passing grades in all subjects. Signed by the director, the student's homeroom teacher, the chairman of the students' committee on education, the secretary of the Komsomol committee (GK VLKSM, 1947, p. 43).

A possible outcome of exerting influence "through the children themselves" emerges from the surviving description of the case of Ksenia L., who committed suicide in 1935 due to her low achievement in mathematics (RONO, 1935). The head of the city school board presented this case as follows:

> As conversations with Komsomol members have revealed, they saw that L. needed help, but took no practical measures to this end, and

> when they did attempt to help her, they did so extremely ineptly. Thus, trying to get L. to improve her performance, the Komsomol organization in April sent two Komsomol members, tenth-graders, to her apartment. Not finding L. at home, the Komsomol members told the apartment representative[3] about her unsatisfactory schoolwork. When Ksenia came home and found out about this, as her mother has related, she became terribly upset and considered herself to have been shamed (RONO, 1935, p. 232).

The head of the city school board found it expedient to confine himself in this case to emphasizing the need for a sensitive understanding of and a solicitous approach toward the students (RONO, 1935, p. 233).

While the struggle against low achievement rates was waged relentlessly, the fight against grade inflation and what was at that time called "liberalism" or even "rotten liberalism" in the assessment of students' knowledge was no less zealous. Teachers were required to make reports about these shortcomings and about the fight against these phenomena. Thus, after reading in a district school board report that "in seventh grade, students have learned to reason logically, to reach correct conclusions, and to solve problems in a fully satisfactory manner," a Communist Party city committee official underlined this sentence in red and wrote ironically: "How good they are!" Nearby, another reader at the party city committee continued in the same vein, adding: "No analysis of shortcomings in the acquisition of mathematical knowledge!" (Vasileostrovsky ONO, 1946, p. 46). Another district school board report for 1946 was phrased in a manner that was more in keeping with the requirements: "The central achievement in the teaching of mathematics this year resides in increased precision and rigidity in the requirements, and the elimination of liberalism in grading" (Krasnogvardeysky ONO, 1946, p. 16).

---

[3]Most city dwellers lived in so-called communal apartments, that is, in apartments that housed several families. Consequently, in each such apartment, one resident was considered responsible for maintaining order, ensuring that bills were paid on time by all the residents, and so on.

At the same time, it was no secret to anyone that grade inflation was a rather natural response to heightened demands:

> This [liberalism in the assessment of students' knowledge] represents a very significant danger under the current circumstances, when teachers are being criticized for having large numbers of failing students and when these teachers' work is not deeply analyzed and mistakes in shortcomings in their work are not discovered. Under such circumstances, naturally, some teachers may conclude that "there's no point in being picky — I'll give them C's and won't get criticized" (LenGorONO, 1946, pp. 79–80).

Analyzing how exams are conducted, the same annual report notes:

> ... there were certain cases ... of envelopes [with problems] being opened prematurely, tests being stolen, and so on. While the tests and exams were taking place, many cases were discovered when teachers — not having provided a high level of instruction during the year — attempted to conceal the shortcomings of their work by inflating grades and helping out the students through leading questions (LenGorONO, 1946, p. 75).

A decree from the minister of education (MP RSFSR, 1948, p. 3) pointed out that "certain school principals ... in chasing after high, nominal indicators of student achievement rates, engage in anti-government practices: they exert pressure on teachers in their assessments of student achievement." Similar observations were heard in the schools as well (School #24, 1952, p. 12). Even when denying the existence of deceptive "anti-government practices" and window-dressing, educators frequently admitted a certain lack of honesty:

> Do we have signs of window-dressing at the school? In the reports that we submit to the school board, we do not stray one iota from the truth. But is this the case for our entire system of working? No. Take the adjustment of D's at the end of the quarter. At the end of the quarter, a mad scramble begins (School #206, 1952, p. 43).

The recommended response to these and similar tendencies was to increase monitoring, to establish the strictest order, and so on.

13 Monitoring in Education

In general, monitoring was seen as a universal method for fighting against various shortcomings, and it was considered natural to explain the existence of such shortcomings as the result of poor monitoring. For example, the Leningrad city school board report for 1937–1938 (LenGorONO, 1938a, p. 19) stated that "due to the liberalism of the principals and the education authorities, visual aids are used unacceptably rarely," and that:

> the state of the notes in students' notebooks, their mathematical correctness, is unsatisfactory due to teachers' extreme lack of attention to this matter and their unwillingness of display proper persistence and firmness.... The influence of left-wing anachronisms in this matter has still not been eradicated (LenGorONO, 1938a, p. 17).

The teacher had to constantly monitor the student, while the principal and the inspector had to constantly monitor the teacher. The principal and the vice principal were each expected to observe hundreds of classes over the course of a year. For example, the positive experience of one school's administration was described in the following terms (IUU, 1954, pp. 49–50):

> 5 to 10 of a teacher's classes are observed in a row and the observations are written down by the principal and vice principal in a separate notebook (50 pages), kept for every teacher individually. Also recorded in these notebooks, in addition to notes about observed lessons, are the results of tests and independent assignments, the results of oral exams administered by the school's principal, assessments of the state of the students' knowledge, assessments of the state of the students' notebooks.... At the end of the inspection, the teacher's work and all of its aspects are discussed in a meeting either with the principal or with the subject committee. Conclusions and suggestions are recorded in the same notebook. After a period of time, an inspection is made to make sure that the suggestions are being carried out. Every teacher is inspected twice a year. At the end of every half-year, an overall evaluation of the teacher's work during the preceding half-year is recorded and then read out at a staff meeting; at the end of the year, these evaluations are used to compile a performance report for the teacher.

In addition, teachers were inspected regularly by mathematics methodologists from district and city offices. The inspection included observing a lesson; reading and analyzing the plan prepared by the teacher for this lesson; examining the class attendance and grades roster; checking how students were questioned and tested by the teacher, and how the questioning was spaced out during the quarter; analyzing students' tests; talking with the principal and vice principal; and examining the principal's and vice principal's notebooks containing notes about observed lessons (IUU, 1951).

Lessons also were visited by inspectors from the district and city boards of education. These inspectors did not always have a strong background in mathematics: one inspector admitted that she had learned how to solve seventh-grade problems in geometry, but that her knowledge went no further since she herself had graduated from school during the civil war and had not solved problems in geometry or trigonometry in school. Nonetheless, she inspected classes, bringing along with her a "knowledgeable person," but clearly believing herself capable of determining whether the teaching was or was not successful (IUU, 1954, pp. 95–96).

A teacher who had been found insufficiently diligent or ill-prepared could be virtually forced to attend professional development courses and could even be fired. But the examination of a teacher's work at a staff meeting could, by itself, have a powerful effect. On the other hand, principals were also blamed for not observing enough lessons or not documenting their visits in a sufficiently detailed fashion, as were inspectors and methodologists who had not been sufficiently assiduous in carrying out their duties (LenGorONO, 1946).

14 The School Atmosphere and Mathematics

Stalin's educational reforms were only one of the measures that transformed the life of the country in the 1930s: other measures included coercive collectivization, which destroyed individual peasant homesteads, and the purges of the Communist Party, which largely destroyed the party that had brought about the October Revolution. Schools during the age of Stalinist terror lived the same life and existed in the same atmosphere as the rest of the country. When two third-graders were careless in copying a sentence from the blackboard — "With today's salaries, foreign workers do not even have enough money to eat and drink" — and left out the word "foreign," thus making the text "politically harmful," a very serious investigation was conducted at the levels of the party's city committee, and although the teacher was largely cleared of any blame, she was advised that some of her texts were poorly structured (LenGorONO, 1946, p. 207). When a biology teacher organized a club that gave answers to anonymous questions about health and sexuality ("is it true that when girls start menstruating, they stop growing?," "what is harmful about smoking?," and so on), the city school board opined that the aim of the club was to distract students from political issues and sent the teacher's file to the NKVD (KGB) (RONO, 1935, pp. 41–43). Mathematics was taught in these same schools, and like all other teachers, mathematics teachers had to participate in all school activities. (The source just cited — district school board reports from 1935, which were used as a basis for quite serious practical decisions — also contains information about a certain Zernova, who was not just the daughter of a country priest, but "does not prove herself in any way in the work of her school. She teaches mathematics and is indifferent to the rest of the life of the school" (RONO, 1935, p. 26).)

And yet, when actual lessons were involved, mathematics was in a special position. The subject was perceived as the foundation of knowledge that was vital to the country, and therefore the most important objective of mathematics lessons was to teach students mathematics.

Learning in general was considered — at least nominally — to be the main task of the student. The Central Committee's special resolution of 1934 warned that overloading students with social–political assignments was unacceptable (Abakumov *et al.*, 1974, pp. 165–166). Naturally, no one had any intention of relaxing the so-called political-ideological training of the students. Yet, for example, when in 1935 at a large conference a teacher described how from November 28 to December 2 her class had daily taken part in events commemorating the first anniversary of the assassination of Kirov, the secretary of the party's Leningrad Regional Committee, she concluded her remarks by saying:

"So if we take into account how much time our kids spend studying and how much time they spend on entertainment, very little time remains for systematically continuing their education" (GK VKP(b), 1935, p. 28). It is quite telling that no one interrupted or corrected her.

Many complaints about the academic workload have survived. A school board inspector related at a meeting how a girl started crying in class because of overwork:

> In order to do the homework assignments that they are given in school, strong students have to spend 5–6 hours working at home. . . . Some usually work at home until 1–2 a.m. every day, others give up working altogether (IUU, 1954, p. 10).

But the content of such academic workloads varied. Typically, even during the period of the so-called "cosmopolitan campaign" of the 1940s–1950s, when praise for all things Russian and criticism of all things foreign became a stringent requirement, and when lessons in virtually all subjects were saturated with materials that promoted this point of view, mathematics was less affected by this than other subjects. Here, it was recommended that Soviet patriotism, pride in Russian research mathematicians, and so on, be cultivated mainly during club meetings. In class, teachers were told to stick to the curriculum (Karp, 2007a).

It is noteworthy that A. P. Kiselev, who before the Revolution had been a member of the rather right-leaning Octobrist Party, and had even been its candidate for a seat in the State Duma, became a kind of icon of teaching (Karp, 2002). Meanwhile, individuals like the aforementioned Leifert or even his collaborator A. R. Kulisher — who was far less aggressive, but nonetheless took part in the "left-wing," "communist" movement in mathematics — both of whom had occupied influential posts during the 1920s, in subsequent years turned out to be out of touch with the time, received prison sentences, and were barred entirely from participating in any active professional life (Karp, 2008).

Of course, neither mathematics teachers nor much more highly placed mathematics educators could entirely avoid taking part in contemporary ideological rituals and procedures, and they certainly could not express unorthodox views openly. A well-known Leningrad mathematics teacher remarked at a school party meeting devoted to ideological education: "Life offers us daily proof of Comrade Stalin's statement that he works better who raises his political–ideological level" (School #24, 1952, p. 30). And then she went on, magniloquently, to tell about how she herself became spiritually enriched by reading Comrade Stalin's works. In the home of Kiselev himself, politics was never mentioned. A. M. Yaglom (2006, p. 442) recalled that in conversations with Kolmogorov, "there was one taboo—political topics." Even so, the lesson in mathematics remained devoted to mathematics.

15 Schools under Stalin: Conclusion

In many aspects, schools under Stalin imitated czarist-era schools. In 1943, single-sex education was brought back, to be abolished only after Stalin's death. In 1940, students started having to pay for school in the higher grades, which was also later revoked (Abakumov *et al.*, 1974, pp. 173–176). And yet, the difference between the two systems of education remained enormous. A serious and in-depth course in mathematics, which had previously been offered to only a few students, now became accessible to millions, and as a result thousands of people could find out about and fall in love with mathematics. Consequently, much greater demands were placed on subject methodology. In the 1960s, the Soviet psychologist Krutetskii (1976) would criticize Western psychology (particularly Thorndike) for its fatalistic faith in special innate abilities without which students were supposed to be incapable of learning high-school algebra. Krutetskii pointed out that what was really at stake here were not innate abilities, but skills acquired in school. Behind his criticism lay the experience of the Soviet system of education, in which most students completed courses in high-school algebra with a relative degree of success.

Soviet education evolved a tradition of highly intensive lessons based on problem solving and involving a high level of mathematical reasoning and proofs. To be sure, these achievements relied in significant ways on what had been created before the Revolution under completely different conditions. Kiselev's textbooks originally had come out ahead of other textbooks in a competitive context. Schools of the Stalin era did not allow for such competition, and therefore the discovery and introduction of new ideas into the curriculum presented difficulties.

Among the methodological ideas of the 1920s that were rejected during the 1930s, many had been popular even before the Revolution, and would subsequently — in the 1960s or later — return to the schools. It is quite another matter, again, that the implementation of methodological reforms during the 1920s not infrequently included semiliterate political propaganda, was conducted with swashbuckling decisiveness, and, most importantly, was generally accompanied by a radical reduction of attention to mathematics.

During the 1930s, when the country intensively began to develop its industry — first and foremost, its military industry — mathematics came to be seen as practically the most important of all subjects. Mathematics also found itself in a privileged position with respect to other subjects thanks to its relative freedom from ideological and other political campaigns. These changes in the subject's status could not but have effects on the success of its teaching.

This success, however, was limited. If during the 1920s those who believed that mathematics was necessary only as an auxiliary and developmental subject tended to exert monopolistic control over Soviet education (although in reality they were not always entirely victorious, due, for example, to an insufficient capacity to carry out administrative monitoring), then in subsequent years it was the model of the theoretical course in mathematics that became hegemonic. Those who had no need for serious theoretical mathematics were not offered anything in its place. But the educational system in principle placed little value on the desires, inclinations, and feelings of individual people.

Daily, all-pervasive monitoring and a "fight for knowledge" in some measure increased the intensiveness of the teacher's labors, facilitating the creation of a school of methodology. At the same time, in a system that depended on checking and monitoring, the growth of fraud, lies, and corruption was virtually inevitable, all of which became particularly apparent later on when, due to a shortage of resources, the regime became less repressive than it was under Stalin.

The system of mathematics education that took shape between the 1930s and the 1950s had great achievements and profound problems. All subsequent developments in Russian mathematics education in one way or another made use of what had been achieved during these years or attempted to reorganize and reform it.

References

Abakumov, A. A., Kuzin, N. P., Puzyrev, F. I., and Litvinov, L. F. (Eds.). (1974). *Narodnoe obrazovanie v SSSR. Sbornik dokumentov. 1917–1973* [*People's Education in the USSR. Collection of Documents*]. Moscow: Pedagogika.

Andronov, I. K. (1967). *Polveka razvitiya shkolnogo matematicheskogo obrazovaniya v SSSR* [*Fifty Years of Development in Mathematics Education in the USSR*]. Moscow: Prosveschenie.

Berezanskaya, E. (1933). *Matematika v sredney shkole. Instruktivno-metodicheskie materialy NKP RSFSR* [*Mathematics in Secondary Schools. Instructional-Methodological Materials. RSFSR Ministry of Education*]. Moscow: Uchpedgiz.

Blonsky, P. P. (1971). *Moi vospominaniya* [*My Memoirs*]. Moscow: Pedagogika.

Blonsky, P. P. (2006). K voprosu o merakh bor'by so shkol'noy neuspevaemost'iu [On methods for fighting against student failure]. *Psikhologiya mladshego shkol'nika*, 616–620. Moscow-Voronezh: RAO.

Chistiakov, I. (1918). Tseli i metody prepodavaniya matematiki v novoy shkole [Aims and methods of teaching mathematics in the new school]. *Matematika v shkole*, *1*(1–2), 2–6.

Demidov, S. S., and Levshin, B. V. (Eds.). (1999). *Delo akademika N.N. Lusina* [*The N.N. Luzin Affair*]. St. Petersburg: RHGI.

Ewing, E. T. (2002). *The Teachers of Stalinism*. New York: Peter Lang.

GK VKP(b) (1935). *Stenograficheskii otchet soveschaniya po voprosu o vypolnenii postanovleniya TSK VKP(b) i SNK SSSR ot 3/09-35 o shkole [Transcript of a conference on the implementation of resolutions concerning schools passed by the Central Committee of the Communist Party and the Council of People's Commissars of the USSR on September 3, 1935].* TSGA IPD SPb, f. 25, op. 11, d. 4.

GK VKP(b) (1936a). *Stenogramma priema pedagogov i pedologov tov. Zhdanovym [Transcript of Comrade Zhdanov's meeting with teachers and pedologists] 25 June 1936.* TSGA IPD SPB, f. 25, op. 2, d. 628.

GK VKP(b). (1936b). *Stenograficheskii otchet soveschaniya uchiteley pri Gorkome po voprosu o vtorogodnichestve i tret'egodnichestve v shkolakh Leningrada [Transcript of a teachers' conference at the city committee devoted to the topic of students who get held back for two and three years].* TSGA IPD SPb, f. 25, op. 11, d. 60.

GK VLKSM (1947). *Stenogramma soveschaniya v otdele shkol Gorkoma VLKSM s direktorami shkol Leningrada po voprosu o tom, kak Akademia pedagogicheskikh nauk pomogaet v rabote shkol [Transcript of a city school board conference, with the participation of the All-Union Leninist Communist Youth League and the principals of Leningrad's schools, on the assistance provided by the Academy of the Pedagogical Sciences in the work of the schools].* TSGA IPD SPb, f. K-881, op. 8, d. 472.

Gratsianskii, I. I. (1924). Materialy o kompleksnom prepodavanii [Materials on complex teaching]. In I. I. Gratsianskii (Ed.), *Matematika v shkole* (pp. 151–154). Leningrad: Izdatel'stvo knizhnogo sektora GUBONO.

Gratsianskii, I. I. (1926). Metodika matematiki i kompleksnaya systema [The method ology of mathematics and the complex system]. In I. I. Gratsianskii (Ed.), *Matematika v shkole*, Vol. 5 (pp. 8–15). Moscow-Leningrad: Gosudarstvennoe izdatel'stvo.

Gurkina, N. K. (2001). *Istoriya obrazovaniya v Rossii* (10–20 veka) [*History of Education in Russia* (X–XX centuries)]. St. Petersburg: SPbGUAP.

Gurvits, Iu. O., and Gangnus, R. V. (1936). *Sistematicheskii kurs geometrii. Uchebnik dlya nepolnoy sredney i sredney shkoly* [*Systematic Course in Geometry. Textbook for Middle and High Schools*]. Moscow: Gosudarstvennoe uchebno-pedagogicheskoe izdatel'stvo.

Gurvits, Yu, Znamenskii, M., and Fridenberg, V. (1929). Analiz priema v vuzy osen'iu 1928 goda. [Analysis of college admissions in the fall of 1928]. *Fizika, khimiya, matematika, tekhnika v trudovoy shkole*, 5, 76–85.

Holmes, L. (1991). *The Kremlin and the Schoolhouse.* Bloomington and Indianapolis: Indiana University Press.

IUU (1939). *Godovoy otchet o rabote fakul'teta za 1938/39 uchebnyi god [Annual report on the work of the department for the 1938–1939 school year].* TSGA, f. 4966, op. 1, d. 10.

IUU (1945). *Dokladnaya zapiska i analizy Instituta usovershenstvovaniya uchiteley (IUU) o sostoyanii prepodavaniya biologii, matematiki, russkogo yazyka i geografii. [Report and analyses of the Leningrad City Institute of Teachers' Continuing Education (ITCE) on the state of instruction in biology, mathematics, Russian language, and geography].* TSGA IPD, f. 25, op. 11, d. 310.

IUU (1949). *Otchety o sostoyanii prepodavaniya i kachestve znanii uchaschikhsya shkol Leningrada po matematike i arifmetike [Reports on the state of instruction and Leningrad student proficiency levels in mathematics and arithmetic].* TSGA SPb, f. 4966, op. 1, d. 340.

IUU (1951). *Protokoly zasedanii kafedry matematiki IUU [Transcripts of Mathematics Department Meetings from the Institute of Teachers' Continuing Education].* TSGA SPb, f. 4966, op. 1, d. 436.

IUU (1952). *Stenograficheskii otchet soveschaniya kafedry matematiki o putiakh osuschestvleniya politekhnicheskogo obucheniya v shkole [Transcript of a mathematics department conference on ways of implementing polytechnical teaching in the schools].* TSGA SPb, f. 4966, op. 2, d. 99.

IUU (1953). *Stenogramma seminara predsedateley raionnykh komissii o prepodavanii algebry v svete politekhnicheskogo obucheniya [Transcript of a seminar of the heads of district committees on teaching algebra in the light of polytechnical education].* TSGA SPb, f. 4966, op, 2, d. 171.

IUU (1954). *Stenogramma seminara inspektorov RONO i GorONO o sisteme prepodavaniya, metodicheskoi rabote i kontrole za rabotoi uchitelya matematiki [Transcript of a seminar of school board and city school board inspectors on the education system, on methodological work, and on monitoring the work of mathematics teachers]*. TSGA SPb, f. 4966, op. 2, d. 239.

Karp, A. (2002). *Klassik real'nogo obrazovaniya* [*Classic of Genuine Education*]. St. Petersburg: SMIO-Press.

Karp, A. (2007a). The cold war in the Soviet school: A case study of mathematics. *European Education*, *38*(4), 23–43.

Karp, A. (2007b). Exams in algebra in Russia: Toward a history of high stakes testing. *International Journal for the History of Mathematics Education*, *2*(1), 39–57.

Karp, A. (2008). "Ne sootvetstvuet sovremennoy postanovke geometrii" (ob odnom epizode iz istorii matematicheskogo obrazovaniya) ["Not how geometry is taught now" (on an episode from the history of mathematics education)]. In V. V. Orlov (Ed.), *Problemy teorii i praktiki obucheniya matematike* (pp. 20–23). SPb.: RGPU.

Karp, A. (2009a). Back to the future: The conservative reform of mathematics education in the Soviet Union during the 1930s–1940s. *International Journal for the History of Mathematics Education*, *4*(1), 65–80.

Karp, A. (2009b). Teaching the mathematically gifted: An attempt at a historical analysis. In R. Leikin, A. Berman, and B. Koichu (Eds.), *Creativity in Mathematics and the Education of Gifted Students* (pp. 11–30). Rotterdam: Sense Publishers.

Kaverin, V. A. (1950). *Dva kapitana* [*Two captains*]. Moscow: Gosudarstvennoe izdatel'stvo khudozhestvennoy literatury.

Kolyagin, Iu. M. (2001) *Russian Schools and Mathematics Education* [*Russkaya shkola i matematicheskoe obrasovanie*]. M: Prosveschenie.

Kommissariat narodnogo prosvescheniya. (1919). *Materialy po reforme schkoly. Primernye programmy po matematike [Materials on school reforms. Model programs in mathematics]*. Izdanie otdela podgotovki uchiteley Komissariata narodnogo prosvescheniya Soiuza kommun Severnoy oblasti.

Korolev, F. F. (1958). *Ocherki po istorii sovetskoy shkoly i pedagogiki. 1917–1920* [*Sketches in the History of Soviet Schools and Pedagogy. 1917–1920*]. Moscow: Izdatel'stvo Akademii pedagogicheskikh nauk RSFSR.

Korolev, F. F., Korneichik, T. D., and Ravkin, Z.I. (1961). *Ocherki po istorii sovetskoy shkoly i pedagogiki. 1921–1931* [*Sketches in the History of Soviet Schools and Pedagogy. 1921–1931*]. Moscow: Izdatel'stvo Akademii pedagogicheskikh nauk RSFSR.

Krasnogvardeysky ONO. (1946). *Otchet Krasnogvardeyskogo rayionnogo otdela narodnogo obrazovaniya o rabote shkol rayiona za 1945–46 [Krasnogvardeysky district school board report on the work of the schools in the district for 1945–46]*. TSGA IPD SPb, f. 25, op. 11, d. 399.

Krogius, V. A., and Popov, N. S. (1928). *Sputnik prepodavatelya matematiki* [*The Mathematics Teacher's Companion*]. Moscow-Leningrad: Gosudarstvennoe izdatel'stvo.

Krutetskii, V. A. (1976). *The psychology of mathematical abilities in schoolchildren*. J. Kilpatrick and I. Wirszup (Eds.). Chicago: University of Chicago Press.

Leifert, L. A. (Ed.). (1931). *Metodika matematiki dlya pedagogicheskikh tekhnikumov* [*The Methodology of Teaching Mathematics. Textbook for Pedagogical Schools*]. Moscow-Leningrad: OGIZ.

LenGorONO (1932). *Otchet LenGorono o rabote shkol za 1931-32 gg [Leningrad city school board report on the work of schools in 1931-32] (part 2)*. TSGA, f. 5039, op. 3, d. 2.

LenGorONO (1934). *Stenogramma disputa po voprosam metodicheskogo rukovodstva shkolami [Transcript of a debate on methodological guidelines for schools]*. TSGA SPb, f. 5039, op. 3, d. 38.

LenGorONO (1936a). *Stenogramma soveschaniya s inspektorami Volodarskogo i Petrogradskogo raionov o prorabotke materialov Dekabr'skogo plenuma TSK VKP (b) i postanovleniya SNK i TSK VKP(b) ot 3 sentyabrya o shkole [Transcript of a meeting with Volodarsky and Petrogradsky District inspectors containing a discussion of the materials of the December plenary meeting of the Central Committee of the Communist Party and the Central Committee resolutions of September 3 concerning schools]*. TSGA SPb, f. 5039, op. 3, d. 95.

LenGorONO (1936b). *Soveschanie s nauchnymi rabotnikami po voprosu postanovki prepodavaniya v shkolakh matematiki, geografii, istorii [Conference with Scientists on the Subject of Teaching Mathematics, Geography, and History in the Schools]*. TSGA SPb, f. 5039, op. 3, d. 140.

LenGorONO (1936c). *Dokladnye zapiski i otchety o sostoyanii prepodavaniya v shkolakh [Reports and briefs on the state of teaching in schools]*. TSGA SPb, f. 5039, op. 3, d. 154.

LenGorONO (1938a). *Dokladnaya zapiska LenGorono ob itogakh uchebno-vospitatel'noy raboty za 1937–38 uchebnyi god v shkolakh Leningrada [Leningrad city school board report on the outcomes of teaching and instruction for the 1937–38 school year in Leningrad schools]*. TSGA SPb, f. 5039, op. 3, d. 201.

LenGorONO (1938b). *Stenogramma soveschaniya uchiteley matematiki [Transcript of mathematics teachers' conference]*. TSGA SPb, f. 5039, op. 3, d. 248.

LenGorONO (1946). *Godovoy otchet Gorodskogo otdela narodnogo obrazovaniya, spravki inspektorov o sostoyanii uchebno-vospitatel'noy raboty shkol i znanii uchaschikhsya g. Leningrada, 1945–46 uchebnyi god [Annual report of the Leningrad city school board and inspectors' reports on the state of teaching and instruction in schools and Leningrad students' proficiency for the 1945–46 school year]*. TSGA IPD, f. 25, op. 11, d. 343.

Lomakin, I. (1919). Po shkolam Moskvy. (Iz otchetov shkolnykh instruktorov) [Around Moscow's Schools (from the Reports of School Supervisors)]. *Narodnoe prosveschenie*, *29*, 17–18.

Matematika. Materialy o sostoianii i rezul'tatakh prepodavaniya v shkolakh Leningrada v 1947/48 uchebnom godu [*Mathematics. Materials on the Conditions and Outcomes of Teaching in Leningrad Schools for the 1947–48 School Year*]. (1948). Leningrad.

Modenov, P. S. (1954). *Sbornik zadach po matematike* [*Problem Book in Mathematics*]. Moscow: Sovetskaya nauka.

MP RSFSR (1948). *Ob itogakh raboty shkol RSFSR v 1947/48 uchebnom godu i ocherednykh zadachakh na 1948/1949 uchebnyi god. Prikaz Ministra prosvescheniya RSFSR [On the outcomes of the work of schools for the 1947–48 school year and the objectives for the 1948–49 school year. Order of the Minister of Education of the RSFSR].* Moscow: MP RSFSR.

OK VKP(b) (1936a). *Dokladnye zapiski, pisma i zayavleniya predstaviteley OK VKP(b) i raionnykh, mestnykh partiinykh i shkol'nykh organov o khode vypolneniya postanovleniya TSK VKP(b) "O pedologicheskikh izvrascheniyakh v sisteme Narkomprosov" i "Ob organizatsii uchebnoy raboty i vnutrennem rasporiadke v shkole" [Reports, letters, and notices of CPSU regional committees and district, local party, and school offices on the implementation of the Central Committee's resolution "On pedagogical perversions in the system of the People's Commissariats of Education" and "On the organization of teaching and internal regulations in the school"].* TSGA IPD SPb, f. 24, op. 11, d. 37.

OK VKP(b) (1936b). *Dokladnye zapiski, otchety, svedeniya i perepiska s Krasnogvardeiskim RONO, OBLONO, GORONO, RK VKP(b) o rabote shkol, tekhnikumov, vypolnenii reshenii TSK VKP(b) [Reports, notices, and correspondence with the Krasnogvardeysky district school board, regional school board, city school board, and CPSU district committee on the work of schools and vocational schools in implementing the Central Committee's resolutions].* TSGA IPD SPb, f. 24, op. 11, d. 38.

OK VKP(b) (1947). *Perepiska s LenGorONO i spravki otdela shkol Obkoma [Correspondence with the Leningrad city school board and reports from the School Department of the Regional Committee].* TSGA IPD SPb, f. 24, op. 2-v, d. 8187.

Otchet o prepodavanii matematiki v shkolakh Leningrada i gubernii [Report on mathematics education in Leningrad and Leningrad province schools] (1926). In I. I. Gratsianskii (Ed.), *Matematika v shkole*, (pp. 166–170). Leningrad: Gosudarstvennoe izdatel'stvo.

Pistrak, M. (1929). Ob osnovnykh voprosakh prepodavaniya matematiki v sredney shkole [On the principal issues of mathematics education in secondary schools]. *Fizika, khimiya, matematika, tekhnika v trudovoy shkole, 5*, 4–6.

Ponomareva, A. I. (1926). Opyt razrabotki kompleksnoy temy "narodnaya svyaz' (pochta) [Developing the complex topic of "communications" (post office)]. In I. I. Gratsianskii (Ed.), *Matematika vshkole, Vol. 5*, (pp. 25–40). Moscow-Leningrad: Gosudarstvennoe izdatel'stvo.

Proekt primernogo plana zaniatii po matematike na pervoy stupeni edinoy trudovoy shkoly-kommuny [Draft for a model course of primary instruction in mathematics at the first stage of the unified labor school and commune] (1918). *Matematika v shkole, 1*(1–2), 38–50.

Ravitch, D. (2000). *Left Back: A Century of Failed School Reforms.* New York-London: Simon and Schuster.

RONO (1935). *Informatsionnye svodki RONO o politiko-moral'nom sostoyanii shkol i perepiske so shkolami po etomu voprosy [District school board reports on the political-moral state of schools and correspondence with schools on this topic].* TSGA SPb, f. 5039, op. 1, d. 15.

Sakharov (1938). Novye programmy po matematike na 1937/1938 uchebnyi god [New Programs in Mathematics for the 1937–38 School Year]. *Matematika v shkole, 1*, 77–82.

Scherbina, K. M. (1938). Kriticheskii obzor programmy sredney shkoly po matematike NKP RSFSR [A critical survey of the RSFSR Ministry of Education secondary school mathematics curriculum]. *Matematika v shkole, 5–6*, 92–104.

School #15 (1921). *Otchety pedagogov o vypolnenii uchebnykh program, svedeniya ob uspevaemosti uchaschikhsya za 1920–1921gg (15 trudovaya shkola) [Teachers' reports on the implementation of teaching curricula, information about student success rates for 1920–21 (Labor School No. 15)].* TSGA SPb, f. 3102, op. 1, d. 270.

School #24 (1952). *Protokoly obschikh sobranii pervichnoy partorganizatsii shkoly #24 [Transcripts from the general meetings of the primary party organization of the school #24].* TSGA IPD SPb, f. 3001, op. 1, d. 7.

School #105 (1954). *Protokoly obschikh sobranii pervichnoy partorganizatsii shkoly #105 [Transcripts from the general meetings of the primary party organization of the school #105].* TSGA IPD SPb, f. 7119, op. 1, d. 13.

School #206 (1952). *Protokoly obschikh sobranii pervichnoy partorganizatsii shkoly #206 [Transcripts from the general meetings of the primary party organization of the school #206].* TSGA IPD SPb, f. 3329, op. 3, d. 5.

Shatsky, S. T. (1929). Moy pedagogicheskii put' [My pedagogical journey]. In S. A. Cherepanov (Ed.), *S. T. Shatsky v ego pedagogicheskikh vyskazyvaniyakh.* Moscow.

Stabil'nye uchebniki po matematike (sredniaya shkola) [Standard textbooks in Mathematics (Secondary Schools)]. (1938). *Matematika v shkole, 4.*

Tolstoy, I. I. (2002). *Memuary (Memoirs).* Moscow: Indrik.

Vasileostrovsky ONO (1946). *Otchet Vasiliostrovskogo Otdela narodnogo obrazovaniya za 1945–46 u.g. [Report of the Vasiliovstrovsky school board for 1945–46].* TSGA IPD SPb, f. 25, op. 11, d. 397.

Volberg, O. (1918a). Matematika v trudovoy shkole [Mathematics in the school of labor]. *Matematika v shkole, 1*(1–2), 9–18.

Volberg, O. (1918b). Neskolko slov ob uchebnike [A few words about the textbook]. *Matematika v shkole, 1*(1–2), 35–37.

Volberg, O. (1919). Dva mirovozzreniya [Two worldviews]. *Narodnoe prosveschenie,* 11–12, 81–86.

Voronets, A. M. (1928). *Metodika matematiki* [*The Methodology of Mathematics*]. Moscow: Rabotnik prosvescheniya.

Venttsel', R. P., and Epstein, G. L. (Eds.). (2007). *Venttsel' E.S. — I. Grekova. K stoletiiu so dnia rozhdeniya. Sbornik* [*On the Centenary of the Birth. Collection*]. M.: Izdatel'skii dom Iunost'.

Yaglom, A. M. (2006). A. N. Kolmogorov glazami cheloveka moego pokoleniya [A. N. Kolmogorov through the eyes of a person of my generation]. In A. Shiriaev (Ed.), *Kolmogorov v vospominaniyakh uchenikov* (pp. 435–441). Moscow: MTSNMO.

3

Toward a History of Mathematics Education Reform in Soviet Schools (1960s–1980s)

Alexander Abramov

1 Introduction

The reform of mathematics education in Soviet schools during the 1960s and 1970s often is linked with the name of Andrey Kolmogorov. This view is well founded. Andrey Nikolayevich Kolmogorov was indeed the recognized leader of the reform. Not a single decision of any importance was made without his involvement. He was both the intellectual force behind the reform and its most active participant. In the rich history of Russian culture, it may be argued, there were only two great personalities who both won worldwide recognition for their achievements in their chosen fields and devoted a considerable part of their lives to the cause of Education. They were Leo Tolstoy and Andrey Kolmogorov.

For this reason, we will begin this sketch of the history of mathematics in Soviet schools with a brief discussion of the following topic: Kolmogorov and schools.

2 Kolmogorov and Schools

Kolmogorov's research in mathematics was published in widely known journals. Collectively, they form a body of work that has been thoroughly analyzed and thoroughly annotated. The fate of his vast pedagogical legacy has been altogether different. His numerous articles are scattered among disparate sources and there are many unpublished texts in the Kolmogorov archives. There is reason to hope that, in the near future, certain obvious gaps in "Kolmogorov studies" will be reduced significantly. After 20 years of studying Kolmogorov's pedagogical works on secondary school education (and he also has works on higher education), I can confirm that this part of his legacy is colossal in scope (a brief overview and an incomplete list of works appear in Abramov, 1988), and extremely rich in ideas and precise observations.

The final 24 years of Kolmogorov's life were devoted to improving mathematics education in Soviet schools. There is a certain mystery here. Why did Kolmogorov, the founder of a great school of mathematics, who by the age of 60 had attained the highest peaks in his discipline, suddenly sharply curtail his work in mathematics and devote himself wholly to education in the schools? There is an element of drama here as well: Kolmogorov's selfless devotion not only was unrecognized, but brought him serious worries. Significantly, notes of regret about his decision, and even notes of disapproval, can be heard from some of his famous students, who remain true to their teacher's memory.

My hypothesis is this:

(1) The simplest way to explain this mystery is to say that Kolmogorov was going through a certain creative crisis — that he felt a lack of fundamental new mathematical ideas. There were psychological grounds for such a crisis, too. Kolmogorov had often said that mathematicians were capable of working to the full extent of their powers until the age of 60, but it is likely, however, that everything was actually much more complicated.

Kolmogorov's genius was by no means limited to mathematics. He was interested in problems connected with everything in the world — nature, humanity, and fields that might seem far from his main interests. His deep works on history, poetics, and linguistics were not accidental. In his half-humorous "Plan for Becoming a Great Man If One Has Enough Will and Energy" (Shiryaev, 2003), which he composed for his 40th birthday (1943), he expressed the intention not only to start writing school-level textbooks at the end of his life, but also to produce a large monograph with the mysterious title "The History of the Forms of Human Thought." It is quite possible that immersing himself in problems connected with schools was an important stage in the realization of some grand design: the school, after all, is a storehouse of all of mankind's "big questions."

(2) This decision, so important for Kolmogorov's personal life, was not made on the spur of the moment. Kolmogorov was a modest person and he never made undue parade of his worldwide fame. At the same time, he knew his own worth, clearly recognized his strength, importance, and responsibility, and believed that it was his moral duty to do everything possible for science, for his homeland, and for humanity.

In this aspect, his decision appears quite logical. After becoming one of the world's most prominent mathematicians by the end of the 1930s, he began communicating his knowledge directly to his students, thus creating — like his teacher Nikolai Luzin — remarkable scientific schools. These schools were based on many ideas and projects that Kolmogorov developed as a teacher of future scientists — as a professor at Moscow State University, who spent a great deal of time working with students. The pyramid "undergraduate students — graduate students — scientists" requires a solid foundation. Therefore, Kolmogorov's involvement in working with talented schoolchildren was a logical next step, as was his subsequent work on improving the teaching of mathematics in schools. Only in this way could the Palace of Mathematics, which Kolmogorov had spent his whole life building, become a completed edifice.

(3) Finally, it should be emphasized that certain key events took place at various stages of Kolmogorov's life and had a particular influence on him. Both Kolmogorov's genius and his personality stem from his childhood, adolescence, and youth. In his articles, letters, and conversations, he often returned to the events of his early life.

First, there was his early childhood. Left without a mother — Maria Kolmogorova died while giving birth to him — Kolmogorov was raised in an atmosphere of love and attention in a wealthy noble family that embraced the best traditions of the Russian intelligentsia, combining a deep interest in culture with respect for work and adherence to democratic principles. Kolmogorov's diligence, inquisitiveness, and talent began to take shape at a very early age. When he was five years old, he made his first mathematical discovery, observing that the sum of consecutive odd numbers is always a perfect square (Kolmogorov, 1988, p. 7).

Second, Kolmogorov's education in E. A. Repman's private gymnasium in Moscow left a very vivid impression on him, over and above the excitement of his first encounter with science. Until the end of his life, Kolmogorov often recalled these years, his school friends, and his teachers. It was during these years, too, that he first began to dream of creating his own school.

Third, there was his time at the university. The atmosphere in Nikolai Luzin's school was the one of scientific exploration and this had an unequivocally beneficial influence on Kolmogorov. His university years witnessed his exceptionally powerful first steps in science. But a very great role was also played by the three years that Kolmogorov spent working as a teacher of mathematics and physics (as well as secretary of the school council, an elected position that he was proud to hold) at the education ministry's Potylikhinsky experimental-model school (Kolmogorov, 1988, p. 9).

In turning to the problems of school education in the 1960s, Kolmogorov was simultaneously repaying a debt to his own teachers and coming back to his old plans and dreams, which had engaged him deeply and sincerely. As he himself said, his attitude toward schools was the one of youthful enthusiasm. Kolmogorov used to express an idea that might shed light on his psychological profile: He believed that every person's development stopped at a certain age — and the lower this "psychological age" was, the more talented was the person. When asked directly: "And how old are you?" Kolmogorov replied: "I am 14." It is important to note, too, that self-conscious reflection, which was a part of his nature, influenced the decisions that he made in the course of the reform. In his youth, he had reflected deeply about his experience as a student and young teacher, and remembered them well.

By the early 1960s, Kolmogorov had already accumulated considerable experience with work on mathematics education. He had published his first article (Kolmogorov, 2007, p. 259) on the popularization of mathematics in 1929. Later on, he continued developing his ideas. Starting in the mid-1930s, on the invitation of the academician O. Yu. Shmidt, who was at that time the editor-in-chief of the *Great Soviet Encyclopedia*, Kolmogorov wrote many encyclopedia articles on mathematics (about 100 in all (Kolmogorov, 2007)). They included the classic article "Mathematics," which contained a holistic view of the development and methodology of this discipline. Many of the texts that Kolmogorov wrote for schools were connected with these encyclopedia articles. It also should be borne in mind that, in his early youth Kolmogorov had studied mathematics on his own using the Brockhaus and Efron Encyclopedia.

Starting in 1935, the year of the first Moscow mathematics olympiad for schoolchildren, Kolmogorov became actively involved in conducting olympiads, and in organizing and working in so-called "mathematics circles" (mathematics clubs for schoolchildren). At the end of the 1950s, he lent his support to the idea of so-called *Youth mathematics schools* (optional evening classes), which had originally been conceived at the Ivanovsky Pedagogical Institute, where the academician A. I. Maltsev — one of Kolmogorov's students — was working at that time (Abramov, 2008).

The present article will deal with the reform of all schools in the Soviet Union. But in order to provide a complete picture of the reform, it is necessary to describe briefly Kolmogorov's activities in facilitating the development of talented children.

In the early 1960s, Kolmogorov helped to organize the All-Russian and then the All-Union mathematics olympiads; he served as chairman of the Jury at the olympiads, and repeatedly traveled to the cities where the final rounds of the olympiads were held. In 1970s, together with I. K. Kikoin, Kolmogorov founded the magazine *Kvant*, remaining until the end of his life the head of the magazine's mathematics department and its deputy editor-in-chief.

But his main concern was with the Physics–Mathematics School that opened under the aegis of Moscow State University in 1963 and now bears Kolmogorov's name. Over the course of 15 years of active work, Kolmogorov taught many vivid courses and delivered numerous lectures at this school, encompassing the whole spectrum of the various fields of mathematics that are accessible to schoolchildren. In 1978, serious illness (Parkinson's disease) severely impaired his speech and limited his mobility. But prior to this, Kolmogorov very frequently visited summer schools to select students for the Physics–Mathematics School, went on camping trips with students, organized literary and musical events. From 1963 until the end of his life, he remained the head of the school's board of advisors. The history of the creation of the Physics–Mathematics School has been described in a recently published book (Abramov, 2008).

It is often forgotten that, by the time that Kolmogorov became the leader of the reform movement, in addition to his experience as a school teacher (during the 1920s), he already had considerable experience as the author of a school textbook. In the years before WWII (1937–1941), Kolmogorov co-authored an algebra textbook with P. S. Alexandrov. Its first part, for grades 6 and 7 (Alexandrov and Kolmogorov, 1940), came out in 1940. Right before the war, the journal *Matematika v shkole* ("Mathematics in the School") published sample chapters from the second part (Alexandrov and Kolmogorov, 1941a,b). An outline of the book's overall plan has survived, which indicates that A. Ya. Khinchin would have also been involved in the project. The war got in the way, however, and when the war ended, work on the textbook was not resumed.

Kolmogorov's earliest sketches for a school curriculum in mathematics date from the years before the war to the late 1940s. It appears that Kolmogorov and Alexandrov played a substantial role in the heated discussions about the teaching of mathematics at the end of the 1930s. The fate of A. P. Kiselev's textbooks, which were later acknowledged to be classics, was far from clear at the end of the 1930s: they were adopted merely on a temporary basis. Kolmogorov once told me that before the war, A. P. Kiselev had visited him and Alexandrov at their "dacha" (vacation cottage) in Komarovka in order to discuss the fate of his textbook and to ask for their support.

Kolmogorov's first programmatic article on school-related issues appeared in 1958 in the newspaper *Trud* (10 December 1958), as part of a discussion of the "Khrushchev education reform" project. Apparently, Kolmogorov made a fundamental decision to begin working with schoolchildren seriously at the end of 1962, shortly before his

60th birthday. In the 31 December edition of the newspaper *Izvestiya*, responding to a question about his plans for the new year, Kolmogorov wrote:

> Let me try, however, to formulate my long-time dreams:
>
> 1. To formulate the general logical foundations of mathematics in a way that would allow them to be taught to fourteen- and fifteen-year-old youngsters.
> 2. To eliminate the distinction between the "rigorous" methods of pure mathematicians and the "non-rigorous" methods of pure reason, employed by applied mathematicians, physicists, engineers.
>
> These two problems are closely linked...

The first part of this plan was largely carried out during the course of the reform.

3 The Pre-History of the Reform

In order to understand the factors that determined the course of the reform of school mathematics education, the initial state of mathematics education in the Soviet Union must be described briefly. On the whole, the situation that had taken shape by the beginning of the 1960s must be characterized as one that was favorable to transformations.

First, by that time, a pronounced atmosphere of respect for education and science had developed in the Soviet Union. In various levels of society, people clearly began to develop an appreciation of the significance of education. Elementary and secondary schools were supplemented by a large-scale network of evening schools, which allowed adults to continue their education and which enjoyed widespread popularity. The government was actively engaged in building and equipping schools, and preparing teachers; the number of publications aimed at students, teachers, and institutions of higher learning continued to grow rapidly. The growing respect for and popularity of science and technology were in large part fueled by the unquestionable achievements of the time: the launching of the first satellite into outer space, the first manned space flight, and so on. It is no secret that the authorities' attention to mathematics and science education was motivated first and foremost by the need to train personnel for the defense industry.

Second, at that time a certain balance was maintained between the number of high-school graduates who were prepared to continue their education in colleges and the number of qualified teachers. During the 1950s, not more than 20% of the students who entered first grade went on to obtain a full secondary school education (10 years). On the whole, teachers who were given positions in high schools were well qualified. A certain number of teachers from pre-Revolutionary gymnasia and "real-schools" had also survived, and they were able to transmit their knowledge to teachers of younger generations. Curiously, strong teachers (and hence, strong students as well) were distributed rather evenly across the country. This was an unexpected consequence of the Gulag system: highly-qualified people were sent to various remote places (one such teacher, for example, was Alexander Solzhenitsyn).

Teacher preparation in pedagogical institutes and continuing education institutes for teachers were aimed specifically at future teachers. The teacher preparation program included a large course in elementary geometry and workshops in solving problems, which — as is well known — forms the substance of mathematics education. Courses in methodology were developed and supplemented with textbooks. In general, it must be said that methodology was worked out quite robustly. Since the mid-1930s, there had been no revolutions in the schools — the same textbooks remained continually in use.

In principle, the decision in the 1930s to create a system of universal education on the model of pre-Revolutionary gymnasia and "real-schools" was a risky move since universal and elite (gymnasium) education are two fundamentally different things. But all of the aforementioned factors made it possible by and large to maintain high standards and to achieve quite decent results.

Thus, computational skills were normally learned by solving problems involving elaborate and numerous computations (above all, problems involving "multilayered fractions"). Problem books in arithmetic contained many intricate problems that developed students' mental agility. Rybkin's famous geometry problem book (Rybkin, 1960) consisted of rather difficult problems, while the course in solid geometry (Kiselev, 1960) actively developed both students' spatial imagination and their logical reasoning ability, necessary for substantiating many theorems and solving problems. Lastly, the development of algorithmic skills and knowledge was facilitated by straightedge-and-compass problems, problems involving the transformation of algebraic and trigonometric expressions, problems involving putting expressions into a form convenient for logarithmic representation, and so on. Not all schoolchildren attained the required levels, but colleges' needs for strong applicants were completely satisfied. Problems on college entrance exams were based on school curricula.

The first attempts at reform began in the late 1950s. The "Khrushchev reform" introduced mandatory eight-year education (replacing mandatory seven-year education). For a brief period (1962–1967), high-school education was expanded to include an 11th grade (instead of ending with 10th grade).

Substantively, the changes were not so great. New textbooks in geometry by I. N. Nikitin (1956) and algebra by A. N. Barsukov (1956) for grades 6–8 were introduced in the late 1950s; and the subject "trigonometry" appeared in the curriculum (Novoselov, 1956). These textbooks were criticized actively, but in their basic conception they differed little from the earlier textbooks.

An attempt to change curricula and textbooks was undertaken in 1962, when an open competition for new mathematics textbooks was announced. The chairman of the panel of judges was B. V. Gnedenko; the chairmen of the committees on arithmetic, algebra, and geometry were Professor V. I. Levin from the Moscow State Pedagogical Institute, the famous algebraist A. G. Kurosh from Moscow State University, and Professor N. F. Chetverukhin.

Eighty six authors' groups participated in the competition. Most of them produced patently weak work. The only textbook that won the top prize and was recommended for large-scale publication was E. S. Kochetkova's and E. S. Kochetkov's textbook in algebra for upper grades (1965). This textbook also introduced elements of calculus.

Several other authors' groups also won recognition in the competition and subsequently played notable roles in the reform. The second prize was awarded to an authors' group comprised of three authors (A. F. Semyonovich, F. F. Nagibin, and R. S. Cherkasov) for a geometry textbook for grades 6–8. Honorable mentions were awarded to V. M. Klopsky and M. I. Yagodovsky for a geometry textbook for grades 9 and 10, as well as to K. S. Barybin. B. Ye. Veits and I. T. Demidov were likewise awarded an honorable mention for a textbook on algebra and beginning calculus for upper grades. The results of the competition were published in the journal *Matematika v shkole* (nos. 1 and 3, 1964).

In 1964, V. G. Boltyansky and I. M. Yaglom's ninth-grade geometry textbook was published in a large edition (Boltyansky and Yaglom, 1964). It introduced students to new topics: "Geometric Transformations" and "Vectors." This fundamentally new textbook was clearly still rough. It drew much criticism from both scientists and teachers, and survived in schools for only two years. It became evident that updating the school course in mathematics was a difficult problem that would require a systematic approach.

The reform was preceded by a broad and substantive discussion, mainly among university teachers. Articles were published in the journals *Matematika v shkole* and *Matematicheskoye prosveschenie* ("Mathematics Education," series no. 2). N. Ya. Vilenkin, A. A. Lyapunov, V. G. Boltyansky, and others actively participated in the discussion.

Although opinions about details differed, mathematicians and college teachers agreed that the course in mathematics had become timed. Substantive suggestions for updating the mathematics curriculum boiled down to the following: it was necessary to introduce elements of calculus and analytic geometry, vector algebra, and geometric transformations into the high-school mathematics program. Methodological articles and pedagogical texts demonstrating different approaches to presenting these new topics began to appear.

4 The Curriculum of 1968

The ideology and principal aims of the reform of school mathematics education were largely determined during the preparation of the new mathematics curriculum, which was approved in 1968. The history of the creation of this curriculum deserves special attention.

It should be noted that the reform affected not only the course in mathematics, but the entire contents of school-level education. In December 1966, the Central Committee of the CPSU and the Council of Ministers passed a resolution that determined school policies for many years to come. At the time, it was customary to prepare for political decisions well ahead of time. By the beginning of 1965, a Central Committee for Developing the Content of School Education was established under the aegis of the USSR Academy of Sciences and the USSR Academy of Pedagogical Sciences, chaired by Academy of Pedagogical Sciences vice president A. I. Markushevich. The choice of chairman could hardly have been better. A. I. Markushevich had a great deal of experience in organizational work (since the end of the 1950s, he had been deputy minister of education), and most importantly, he was a highly cultured person, a well-known mathematician — a specialist in complex analysis and a Moscow State University professor — and a wonderful author and popularizer. Markushevich was highly respected both in academic circles and in the educational system. With respect to the reform, it was also significant that Markushevich and Kolmogorov were linked by long-standing relations of mutual respect.

Within the Central Committee, subject committees were formed. Like the mathematics committee, which was chaired by Kolmogorov, the other subject committees were chaired by well-known scientists–academicians: I. K. Kikoin (physics), M. V. Nechkina (history), D. D. Blagoy (literature), and so on. Such participation by major scholars facilitated the aims of the reform: freeing the courses from archaic, second-rate materials, and making them more rigorously scientific (this was motivated, of course, by a wish to accelerate scientific–technological progress and to surpass the USSR's principal Cold War adversary, the United States).

The first document of the reform — "The Scope of Knowledge in Mathematics for the Eight-Year School" (*Matematika v shkole,* 1965, no. 2) — was prepared by members of the Committee on Mathematics Education at the mathematics division of the USSR Academy of Sciences (I. M. Gelfand, A. N. Kolmogorov, A. I. Markushevich,

A. D. Myshkis, D. K. Faddeev, and I. M. Yaglom), i.e., without the participation of methodologists or teachers.

By contrast with ordinary programs, this text did not contain a detailed presentation of topics arranged by grades and subjects in some determined sequence. Rather, in an extremely concise fashion, it described the key ideas that students were required to absorb by the end of their eight years of schooling. The decision to present the program in such a brief form made it possible for people with widely differing views to agree on a common position. Arguments about the contents of school-level education can go on indefinitely. In order to avoid this, it is necessary to agree on key principles, which was the aim of "The Scope of Knowledge." It was expected that a broad discussion would follow and that a detailed program would then be formulated.

Authors' texts have survived that show that the main work on preparing the section on "Arithmetic and Algebra" was done by Kolmogorov; the section on "Geometry" was written by I. M. Yaglom. Drafts for a "Scope of Knowledge in Mathematics for Grades 9–10" have survived in Kolmogorov's archives; these were supposed to be published during the same year. But this plan was changed due to the more active role assumed by A. I. Markushevich's committee.

A "General Explanatory Brief on the Draft of the Curriculum and Programs for Secondary Schools" was published in 1965, followed by curricula in all subjects, including mathematics. But work continued for a long time to come. A pamphlet with the text of the mathematics curriculum was published in 1966 in an edition of 4000 copies (Mathematics Curricula, 1966), which were distributed in all the major cities of the Soviet Union. The pamphlet was discussed very widely, with a great number of people voicing their opinions, which were mainly positive. After some not very substantial revisions, the draft was published for a large-scale audience in *Matematika v shkole* (1967, no. 1), and only at the beginning of 1968 and after another discussion did a final document appear with the endorsement of the Ministry of Education (*Matematika v shkole*, 1968, no. 2).

Thus, work on the curriculum took about three years, which was accompanied by broad discussions, and largely reflected the consensus of the professional community.

The group of individuals who developed the curriculum included scientists, methodologists, and teachers. The mathematicians were represented by V. G. Boltyansky, Kolmogorov, A. I. Markushevich, and I. M. Yaglom. The methodologists were represented by G. G. Maslova, the head of the mathematics education laboratory at the Scientific Research Institute on Educational Content and Methods under the aegis of the USSR Academy of Pedagogical Sciences, as well as this laboratory's members, Yu. N. Makarychev, K. I. Neshkov, A. D. Semushin, and A. A. Shershevsky (one of the best mathematics teachers in Moscow). A. I. Fetisov was a well-known methodologist and author of manuals and problem books in geometry. The prefatory note to the curriculum stated: "The final draft of the explanatory note was completed by A. N. Kolmogorov, A. I. Markushevich (introduction, arithmetic, algebra, and beginning calculus), and I. M. Yaglom (geometry)."

The 1968 curriculum provided for a radical reform of the existing course in mathematics.

The introduction of a series of major new topics significantly expanded the range of information covered; these included elements of calculus, geometrical transformations, vectors and coordinates. Students also were to be given a substantive introduction to the axiomatic method. All of these served the central aim of the curriculum, which was to enrich the course in mathematics with ideas that had become significant in an age of accelerating scientific-technological progress as elements of a common culture. Another important goal was to increase the logical purity of the exposition.

A substantial expansion of the range of subjects and ideas covered in school could be achieved only by allotting time to them in classes. This meant that certain traditional themes and topics had to be abandoned. In this connection, the following decisions were made:

1. The elementary school curriculum in mathematics was shortened from four to three years, while its overall substance was preserved intact.
2. "Arithmetic," as a separate subject, was eliminated. For grades 1–5, a single subject — "Mathematics" — was introduced. It contained

elements of arithmetic as well as preparatory materials for classes in algebra and geometry.

3. While the ideas were raised to a higher level, the level of technical skills that average students were expected to master was lowered, as was the level of difficulty of the problems that average students were given. Different requirements were introduced for different students through the creation of elective classes to be chosen by the students themselves in accordance with their interests, inclinations, and abilities.
4. The topic "complex numbers" was eliminated from the program. The study of elementary probability theory and mathematical statistics had to be abandoned due to the shortage of class time and the lack of sufficiently prepared teachers who had real experience with these subjects.
5. The list of traditionally-studied isolated facts and properties (trigonometric identities, the properties of chords and tangents, and so on) was reduced substantially.
6. The presentation of traditional topics was made more concise and simple through the effective use of new methods (for example, complicated derivations of the formulas for the volume of the pyramid and the sphere, and the area of the sphere, were to be simplified substantially by applying the concept of the integral).

The new curriculum exposed students to elementary set theory and mathematical logic early on. But on the whole, this innovation was moderate by comparison with the reforms that were taking place at the same time in France or Belgium. As the explanatory note that accompanied the curriculum emphasized: "The curriculum approaches the introduction of the concepts and terminology of set theory and mathematical logic with caution. The possibility of using them in schools on a broader scale is still under discussion."

One of the important general principles of the reform was the need to establish a more precise and complete system of notation and exposition for mathematical texts. Kolmogorov connected this directly with the explosive growth in information technology that was expected to take place in the future. Working with machines requires precision and familiarity with working with symbols.

The adoption of the 1968 curriculum opened the door for work on textbooks that could implement the reform's ideas. But existing textbooks were already being revised by the mid-1960s and some obvious shortcomings were being eliminated (Kolmogorov, 1966a, 1967a,b). At the same time, large-scale work was underway on elucidating the ideas of the reform and providing a preliminary presentation of the new topics. *Matematika v shkole* began to publish a series of articles by Kolmogorov and others, aimed at popularizing the new ideas. The publishing houses "Mir" and "Prosveschenie" published a number of books and pamphlets on the "new school mathematics" (Markushevich, Maslova, and Cherkasov, 1978) including translations of foreign texts and textbooks (Moise and Downs, 1968; Doneddu, 1979).

5 The Implementation of the Reforms

The reform involved a large amount of varied work on the territory of an enormous country whose population spoke many languages. In addition, the cultural map of the USSR was highly heterogeneous — there were obvious differences, for example, between the rural schools of Central Asia and the urban schools of the Baltic republics. In order to carry out the reforms, an effective system of management had to be created.

Political decisions were made at the top and passed down to lower governing bodies to be carried out in the school departments of specific party organizations: the hierarchy descended from the Central Committee of the CPSU to the central committees of the republics to the regional committees to the city committees to the district committees. At each of these levels, appropriate goals were set and appropriate decisions were made. It may be said that the role played by the organs of the party was a legislative one. The executive role was played by the educational organs of the Soviet government: the USSR Ministry of Education, the ministries of the republics, the regional school board, the city school board, and the district school board.

The creation of the USSR Ministry of Education in 1966 — prior to which point there had only been ministries in the separate republics — was largely motivated by the need to coordinate the implementation of the reforms. Minister of Education M. A. Prokofiev was a member of the existing establishment, but as a serious scientist (specialist in chemistry, member of the USSR Academy of Sciences) and a genuine activist in the field of education, he remained on mutually respectful terms with Kolmogorov, Kikoin, and other leaders of the reforms. He resigned in 1984, refusing to implement the newly formulated program of bringing informatics into the educational system, considering it unrealistic. After his resignation, he actively promoted the idea of making schools more differentiated, which was rejected by the leadership of the country at that time. He left a testament of sorts in his small book, *Postwar Schools in Russia* (Prokofiev, 1997). In a private conversation, M. A. Prokofiev told me that in the Politburo he had always been supported by Minister of Defense D. F. Ustinov, who understood the significance of schools for the modern army perfectly.

The key decisions (assessing the state of affairs, recommending textbooks, and so on) were made at regularly scheduled Ministry of Education board meetings. An important role was played by the inspectorate of the Ministry of Education, which regularly organized comprehensive inspections across the country.

The system for preparing teachers was also structured hierarchically: from central institutes in the republics to regional continuing education institutes for teachers to district offices to methodological associations in the schools. Regular courses for methodologists from the republics and RSFSR (Russian Federation) methodologists in mathematics were conducted for a number of years in Moscow, at the Central Continuing Education Institute for Teachers. The authors of new textbooks that were to go into use on the first day of school would give lectures; then, the same materials would be presented to teachers — somewhat less cogently, perhaps — during summer and winter courses in regional centers and major cities in the republics.

Responsibility for the scientific side of the reforms — analyzing students' knowledge, analyzing programs and textbooks, developing pedagogical and analytical materials, and so on — was given to the

USSR Academy of Pedagogical Sciences, which had been formed on the basis of the RSFSR Academy of Pedagogical Sciences, also in 1966. The Academy of Pedagogical Sciences communicated and collaborated with pedagogical institutes in all of the republics.

The Academy of Pedagogical Sciences' Scientific Research Institute on Educational Content and Methods oversaw the development of new trial textbooks. In mathematics, this work (making trips to districts where experimental textbooks were being used, analyzing the results, conducting tests, engaging in methodological work with teachers) was carried out by the mathematics education laboratory at the same research institute. The head of the laboratory was G. G. Maslova. Four districts were selected for testing out experimental textbooks: the Tosno district in the Leningrad region; the Beloyarsk district in the Sverdlovsk region; the Suzdal district in the Vladimir region; and the city of Sevastopol. All schools in these districts used two competing textbooks from the late 1960s until the mid-1970s, at which point a final selection of textbooks was made.

The Research Institute on Educational Content and Methods had a strong graduate school. During the 1970s, a large number of pedagogical doctoral dissertations defended at the graduate school dealt with problems connected with the reform of the school mathematics curriculum.

When work on the curricula was completed in 1970, the Central Committee on Content Development was dissolved; its work as a whole was approved at a joint meeting of the presidiums of the USSR Academy of Sciences and the USSR Academy of Pedagogical Sciences, chaired by Academy of Sciences President M. V. Keldysh. A new Scientific Methodological Council — made up of different subject committees — was established at the Ministry of Education in order to oversee the publication of the new textbooks and methodological manuals. Kolmogorov was appointed head of the mathematics committee in 1970. In 1980, he was replaced by the academician A. D. Aleksandrov. The Scientific Methodological Council remained in existence until 1991, i.e., until the collapse of the Soviet Union. Subsequently, it was reorganized into a council of experts, effectively remaining what it had always been, until finally being dissolved in early 2003.

The members of the Scientific Methodological Council were famous mathematicians, methodologists, and teachers. When manuscripts were discussed, two or three principal reviewers would make presentations, summing up the numerous responses to the textbooks received from pedagogical institutes in different republics and regional continuing education institutes for teachers.

Meetings took place approximately every three or four weeks (depending on the number of manuscripts that had to be examined). The discussions were chaired by Kolmogorov, who always familiarized himself with the manuscripts beforehand. Kolmogorov possessed the rare talent of seeing the book in front of him as a whole: after looking through it rather quickly, he would locate what was most essential in it, whether this was an ineffective approach to a subject, obvious mistakes, or, on the contrary, some positive characteristic.

The textbooks and methodological manuals were edited at the mathematics division of the publishing house "Prosveschenie," at that time the largest publishing house in the world. The head of the publishing house, D. D. Zuev, took an active interest in the problematic aspects of school textbooks, created a special committee at the publishing house to work on them, and published 20 volumes of articles on "The Problematic Aspects of School Textbooks." Educational–methodological kits began to be published: these contained not only the textbook itself, but also a manual for teachers and educational materials (tests and quizzes). After work on the textbooks was finished, "Prosveschenie" began publishing a series entitled "The Mathematics Teacher's Library."

As a rule, final decisions about revising the textbooks would be made at the last moment, which made the editorial-and-publication process extremely difficult: new editions of four million copies of a textbook had to be made available by the beginning of the school year. Nonetheless, first editions contained relatively few major flaws (not counting misprints and mistakes in answers to problems).

Publishing houses in the different republics that specialized in education would translate the textbooks into the different languages spoken in the Soviet Union. They would also publish methodological literature by local authors. Relevant and up-to-date information would be published in the journal *Matematika v shkole* (for example, Kolmogorov and Semyonovich, 1970; Kolmogorov and Shvartsburd, 1975).

6 Elective Classes

As a mathematician, Kolmogorov was distinguished by astonishing scientific boldness. He took up problems that seemed unapproachable and managed to solve many of them. The problem that Kolmogorov set before himself in reforming mathematics education was also distinguished by the audacity of its conception. His premise was that the potential of the individual student and the potential of the education system were both high. Therefore, a rather high general level could realistically be attained if education was structured with intelligence and skill. Consequently, the level that the reforms aimed at was substantially higher than the level that was typical of virtually all other countries.

The first phase of the reform would be devoted to finding simple and succinct forms of presentation, a goal that was expressed in Kolmogorov's intention "to formulate the logical foundations of mathematics in a way that a teenager could understand."

But there was also another side to things. What kind of educational system could most effectively develop children's interests, inclinations, and abilities? This second problem had great significance for the government, since the government was particularly interested in finding a means to prepare large numbers of highly qualified experts.

The difficulty resided in ideological constraints: the misleading concept of the "uniformity of the school" (effectively, the idea that education meant the same thing to everyone) made it impossible to introduce differentiations into schools. A democratic solution to this problem was found: it consisted in offering students classes to choose for themselves, i.e., elective classes. Apparently, as the following documents show, this idea was first proposed by Kolmogorov:

> Letter from A. N. Kolmogorov to A. I. Markushevich
>
> (December 29, 1964)
>
> Dear Aleksey Ivanovich!
>
> Please forgive me for the way in which I expressed myself during our recent conversation.

In essence, however, creating possibilities for additional lessons in mathematics and physics in most of the schools in the country remains a very necessary goal if we wish to make further studies in these disciplines and in modern technology genuinely accessible to students. If we expand the programs in all schools by introducing integral calculus, etc., we will thereby also expand the program of college entrance exams. But in most schools, with mediocre teachers and six hours of classes in grades 9–10, students will assimilate the expanded program even worse than they absorb the current curriculum, and naturally, they will not be able to enter any college at all.

Placing all bets on mathematics circles and youth mathematical schools does not seem to me very promising.

But perhaps it is possible, without going against the "uniform school" dogma, to provide time for elective classes in the lesson plans for grades 9–10 (for example, three in ninth grade and six in tenth grade), with the school being obligated to organize them in accordance with the population's wishes. They may even be classes in drawing and radio technology, but they may also be classes in biology and the foundations of evolution, in foreign languages, or in mathematics and physics. What is important is that these will be hours allocated for classes during the entire year, and not just practice internships for some number of work days (now, I believe, 36 days in ninth grade and 12 days in tenth grade) for acquiring expertise and work qualifications.

I have just visited the neighboring Bolshevsky school no. 3. The youngsters take their qualifications as radio technicians quite seriously, but a large number of them would be enthusiastic about three or four hours per week of additional classes in mathematics. The parents, once they find out about such a possibility, would of course want their children to study mathematics, or technical drawing, or foreign languages, and would themselves find expert teachers.

I think that, in altered form, all of this also applies to good schools in state farms, although perhaps not to secondary schools in every backwater village.

There is another question concerning which I should like to know your opinion.

> I can understand the reluctance to expand the network of physics–mathematics schools such as our boarding school to a very large scale. But it is not clear to me whether the people making these decisions realize just how microscopic this whole initiative is, even if it is seen just as an experiment. Responsible government workers, ministers, and deputy ministers meet with university presidents for serious discussions, the television broadcasts my lectures, etc. Yet the idea of selecting 180 students from 40 regions is completely absurd if we believe that we will be able to identify and locate the talents hidden among "the people."
>
> Along what channels should one try to promote the idea that even experimental work must be done on a somewhat larger scale?

Yours, A. Kolmogorov

A. I. Markushevich's response indicates that he too appreciated the absurdity of the "uniformity principle" in Soviet schools. As he wrote:

> In my view of physics–mathematics schools attached to universities as special points within the process as a whole, I apparently have no disagreement with you, Andrey Nikolayevich. But, by contrast with you, I attribute greater importance to schools that continue to prepare computer programmers. After all, it was supremely important to break the bleak bureaucratic monotony of our pre-reform secondary schools, which considered it a virtue to give all of our schoolchildren one and the same thing.

The idea of elective classes developed rapidly. This may be explained, on the one hand, by the fact that organizational problems met with an effective and timely solution. A resolution passed by the CPSU Central Committee in 1966 provided for allotting a certain amount of school time to elective classes and for paying teachers to conduct them. On the other hand, the experience of working with mathematics circles and schools specializing in mathematics that had been accumulated by that time, and most importantly, the involvement of highly qualified authors, made it possible to develop compact elective classes very quickly.

By 1970, the first textbooks for elective classes were completed. They were further developed in the following years (the laboratory for applied mathematics, headed by S. I. Shvartsburd, took charge of organizing the project as a whole, and a particularly prominent role was played by V. V. Firsov, who at that time was one of the laboratory's senior researchers).

For the 1968 curriculum, Kolmogorov had written a special note on elective classes. He proposed creating a course of "Additional Chapters," which would be conceptually connected with the general course. This idea did not take hold. Programs for 17-h and 34-h classes won more support, as did preparatory classes for competitive examinations.

In the late 1960s and early 1970s, the following manual for elective classes was published: "Additional Chapters for the Course in Mathematics" for grades 7 and 8 (Sikorsky, 1969) and grades 9 and 10 (Additional chapters, 1970). In 1978 and 1980, "Selected questions of Mathematics" (Bokovnev and Shvartsburd, 1978; Firsov, 1980) were published. The courses found in these and certain other books are listed below:

V. G. Boltyansky and G. G. Levitas, "The Divisibility of Numbers and Prime Numbers"
R. S. Guter, "Number Systems and the Arithmetic Foundations of Computer Operations"
N. Ya. Vilenkin, "Elements of Set Theory"
I. M. Gelfand, Ye. G. Glagoleva, and A. A. Kirillov, "The Coordinate Method"
I. M. Gelfand, Ye. G. Glagoleva, and E. E. Shnoll, "Functions and Graphs"
K. P. Sikorsky, "Solutions to Problems for the General Course"
A. N. Zemlyakov, "Symmetry"
I. L. Nikolskaya, "Elements of Mathematical Logic"
A. N. Zemlyakov, "Sets on the Coordinate Plane"
N. Ya. Vilenkin, "Infinite Sets"
N. Ya. Vilenkin and A. G. Mordkovich, "Differential Equations"
A. A. Egorov and G. V. Dorofeyev, "Complex Numbers and Polynomials"
A. M. Abramov and A. N. Zemlyakov, "Elements of Spherical Geometry"

In practice, elective classes continued to be developed rather actively throughout the 1970s. But the lack of special measures for preparing teachers for them and the reduction in the number of hours allocated for mathematics held back their development. Although no exact statistics exist, there are reasons to believe that gradually the hours that had been originally intended for elective classes came to be used for preparing students for competitive exams. By the early 1990s, elective classes had dissolved within the school curriculum and ceased to exist.

A mathematics correspondence school was created in 1964 under the aegis of Moscow State University on the initiative of I. M. Gelfand, with the support of I. G. Petrovsky (the rector of Moscow State University). This was a major event — a fundamentally new form of schooling. A system of entrance exams was worked out, and even more importantly, an outstanding system of assignments for students was developed as well. The organization of the school was original and quite democratic; over the course of a two-year program, students were required to complete about 20 substantial assignments. Students' work was checked (and corrected) on a volunteer basis by undergraduates at the mathematics department of Moscow State University: every undergraduate oversaw 10 students, and the work of every 10 undergraduates was monitored by a supervisor — an upperclassman or a graduate student at the mathematics department. The mathematics correspondence school exists to this day (and now encompasses multiple subjects). About 200,000 schoolchildren from many cities and towns have graduated from it; many of them went on to enroll in various colleges.

7 Mathematics 1–5[1]

Before the reforms, students in grades 1–5 had a class called "Arithmetic," which included very minor sections in geometry that dealt mainly with formulas for areas and volumes, and units of

[1] This section and a few following will deal with specific textbooks. The literature review is provided in Shtokalo (1975). Readers can also opt to consult Kolyagin (2001).

measurements — the main motive here was a wish to diversify problems and to give them a practical meaning. The decision to name the new class "Mathematics" reflected those fundamental changes which the reforms had introduced into the education of children between the ages of 7 and 12.

The creators of the new curriculum and the authors of the new textbooks pursued two basic goals: (1) to present the traditional part of the course in a substantially more condensed fashion, including covering a number of topics earlier than before; (2) to include a number of new topics in preparation for classes in algebra and geometry in grades 6–8 — to this end, a number of topics were included in the curriculum for grades 1–5 that had been covered previously in grades 5 and 6.

The most fundamental change was the shortening of elementary school education from four to three years. The contents remained largely what it had been previously: the objective was to study natural numbers, to carry out operations using natural numbers, and to solve easy problems in arithmetic. The principal innovation was the appearance of letter notation and a basic idea of equations. Naturally, the loss of one year of schooling meant that standards for students' computational skills had to be lowered; word problems were made easier as well. The geometrical material was somewhat expanded — students studied the simplest figures and elementary straightedge-and-compass constructions.

After a review, a textbook by M. I. Moro *et al.* was selected for grades 1–3. Until the 1990s, it remained the only textbook in use. In the 1990s, the monopoly was abolished, but this textbook is still used to this day, along with others. A. G. Pchelko, the author of a previously used textbook, contributed to the first editions of this textbook (Moro, Bantova, and Beltiukova, 1968, 1969, 1970), thus helping to provide some continuity between the new curriculum and what had preceded it.

The class "Mathematics 4–5" (Vilenkin *et al.*, 1968, 1969) radically altered the traditional curriculum.

1. The concept of "set" and operations on sets ("intersection" and "union") were explicitly introduced. This terminology and notation was actively employed at subsequent stages of education.

2. In the sections on arithmetic, fractions and negative numbers were introduced earlier than they had been before. The level of difficulty of word problems was lowered. (Traditionally, the following scheme was employed in Soviet and Russian schools: the full solution to a word problem had to include a clearly written out sequence of questions posed by the student, and the calculations required to answer them. Traditional problem books contained extremely involved problems whose solution involved answering 6–10 different questions. Problems in the "reformed" textbook were usually shortened to 2–3 questions per problem.)
3. Explicit algebraization gave the new program a rather revolutionary character. Letter notation, formulas, simple (linear) algebraic equations, and corresponding problems were actively used. All of this was fundamentally new — previously, elements of algebra had first appeared only in grade 6.
4. The list of geometric topics was considerably expanded. These were distributed throughout the entire course. Students were taught coordinates on the line and in the plane. Elementary straightedge-and-compass and protractor problems were solved regularly. An important innovation was the concept of axial symmetry and point symmetry as well as of rotation. The concept of congruent figures was introduced (as a required part of the course). All of this created a foundation for the systematic course in geometry that would begin in sixth grade, in which geometrical transformations played a very important role.
5. A certain lowering in the problems' level of difficulty was compensated for by the inclusion of additional problems with higher levels of difficulty, aimed at developing students' inventiveness.

The textbook "Mathematics 4–5" occupies a special place among all the textbooks that were produced in the course of the reform of mathematics education: it had the calmest, or perhaps the happiest, fate. This conclusion is warranted not only by its longevity: 40 years later the textbook is still used in schools. By contrast with other textbooks, "Mathematics 4–5" was subjected to virtually no criticism either from above or from below. Probably the only shortcoming that teachers saw in it was that it contained too few arithmetical word problems.

I see two basic reasons for its success. First, the trial run of "Mathematics 4–5" lasted longer than the trial runs of other textbooks — four years. This made it possible to analyze its virtues and shortcomings calmly, and to go through several rounds of revisions.

Second (and most importantly), the group of authors who wrote the textbook was well-balanced. N. Ya. Vilenkin and A. I. Markushevich — the textbooks' editor — were the mathematicians among them, but in addition to being major mathematicians, they possessed the intuition of good methodologists and had literary talent. K. I. Neshkov was an exceptionally conscientious and highly talented teacher and methodologist. S. I. Shvartsburd was also a very experienced teacher, who had founded schools specializing in mathematics during the 1950s. He turned out to be a good mediator in discussions that took place among the authors, who were all very different people.

During the 1970s–1980s, the textbook went through rather minor changes. The most notable of them was the "eradication" of set theoretical terminology and notation following the events of 1978–1979 (see below) — although the discussion between Kolmogorov and Vilenkin, who was against introducing the term "congruence," dated back to 1972 (*Matematika v shkole*, 1972, no. 5).

In a competition in 1987–1988, the textbook by Vilenkin *et al.* retained its position, although a new manual by Nurk and Telgmaa (1988) also was introduced. A more significant and consequential event was the beginning of work on a new textbook in arithmetic by a working group led by the academician S. M. Nikolsky (Nikolsky *et al.*, 1988): the authors' main goal was to reestablish arithmetic as the core of the middle school mathematics curriculum.

8 Geometry 6–10

Traditionally, since the 1930s, a systematic course in geometry has been taught in Russia (USSR) from sixth to tenth grade. The reorganization of this course became the single most difficult problem that arose during the reforms. One of the reasons for this was the traditional difficulty of studying geometry at the elementary level, which has even earned a special designation: "the problem of the first lessons in geometry in grade 6." The problem stems from the fact that the deductive style of exposition — something fundamentally new for schoolchildren — requires overcoming both psychological and epistemic difficulties. Schoolchildren do not understand why one must prove things that are obvious. It is also not clear to them why obvious assertions must be proven by using other assertions that are equally obvious.

At the same time, the changes proposed by the new curriculum in this instance were of the most revolutionary nature. There was a great quantity of new materials. Both in terms of its substance and its methodology, this course possessed features that were fundamentally new.

The competition of 1964 was won by a group of authors which included F. F. Nagibin, professor at the Kirov Pedagogical Institute; A. F. Semyonovich, associate professor (and subsequently full professor) at the Cherkassk Pedagogical Institute; and R. S. Cherkasov, professor and chairman of the mathematics teaching methodology department and the Moscow Municipal Pedagogical Institute, and for many years the editor of the journal *Matematika v shkole.* The group was headed by Kolmogorov, who became the co-author and editor of the textbook. His decision to become so involved was based on the extreme time constraints under which the authors had to work. In keeping with the government-mandated schedule for the transition from the old textbooks to the new ones, the first editions of the new textbooks had to meet certain deadlines and it was thus necessary to give the group of authors help in their work.

A first, experimental edition of a textbook in geometry for grade 6 (Kolmogorov *et al.*, 1970) appeared in 1970 (the textbooks "Geometry 7" and "Geometry 8" came out in 1971 and 1972, respectively (Kolmogorov *et al.*, 1971, 1972a,b)). The textbook was put to use on an experimental basis for a three-year trial, which immediately brought to light major problems stemming from the novelty of the theory and the practice for the students, an obviously overloaded curriculum, the teachers' lack of methodological experience, and the novelty of the material for the teachers.

The textbook for grade 6 was substantially reworked, and by 1972 it was introduced in schools across the Soviet Union. The new version of the course in geometry for grades 6–8, which consisted of three textbooks that appeared during the years 1972–1974 (Kolmogorov *et al.*, 1972a,b, 1973a,b, 1974a,b), was distinguished from the earlier, experimental edition by its more systematic approach. The authors sought to correct the numerous flaws that had come to light during the trial run.

Kolmogorov formulated the ideology of this course in the following way (Gusev *et al.*, 1972, p. 7):

> The new course in geometry for grades 6–8 is substantially different from the traditional one. The new textbook in geometry for the eight-year school includes the following changes:
>
> 1. The concept of geometrical figures as sets of points is consistently promoted.
> 2. It is made completely clear (already in sixth grade) that geometry inevitably makes use of certain fundamental concepts that have no obvious definitions, and that these concepts must be used to define precisely all other geometrical concepts.
> 3. The textbook systematically develops the concept of "geometrical transformations" as one-to-one mappings of the entire plane (and later, of all of space) onto itself. In sixth grade, this pertains to "motions" of the plane (in contemporary mathematical language, "isometries"). In seventh grade, students examine similarity transformations, in particular, dilations.
> 4. The textbook gradually prepares materials for understanding the possibility of various "geometries" that are non-Euclidean (such as that of Lobachevsky) or that contain Euclidean geometry as a particular case (such as notion of a "metric space"), which are prepared already in sixth grade through an examination of the basic properties of distances.
> 5. In grade 7, students are introduced to the notion of vector, which is then systematically used in the upper grades and in physics classes.

6. In grade 8, students study the trigonometric functions of angles from $-180°$ to $180°$.

Two more items should be added to this brief list of innovations:

(7) The new program introduced elementary space geometry — a special chapter in the eighth-grade textbook was devoted to this topic.
(8) Students were introduced to the coordinate method in geometry (the equations of the straight line and the circle; elementary problems).

While developing the new course, Kolmogorov paid particular attention to the rigor of the definitions. He believed that the ability to work with rigorous definitions was an absolutely indispensible part of the general skill set of every educated person. In view of the difficulty of some proofs, it is not possible to prove every proposition in a school-level textbook; however, it is important in such cases explicitly to indicate "unproven assertions," whose demonstrable meaning must be convincingly illustrated. But in the case of definitions — and in the formulation of propositions — precision and purity must be observed.

These principles were consistently promoted. The system of exposition that was adopted by the final version of "Geometry 6–8" (Kolmogorov *et al.*, 1979) may be considered flawless in terms of its logical underpinnings. Preference also was given to a precise system of notation and an explicit and comprehensive approach to writing out the solutions of problems.

While working on the textbook, in 1970, Kolmogorov proposed a new, original axiomatization of Euclidean geometry. It was impossible to adhere to the stated principles for structuring the textbook without a clearly articulated system of axioms and basic concepts. Kolmogorov's axiomatics took the following approach (in the 1974 edition of "Geometry 8").

The concepts "point," "straight line," and "distance" are posited as fundamental (undefined) notions. A plane is a set of points in which subsets ("straight lines") are distinguished, "distances" are defined, and the following axioms hold:

I. **Axioms of incidence**

I_1. Each straight line is a set of points.

I_2. For any two points, there exists one and only one straight line that contains them both.

I_3. There exists at least one straight line; every straight line contains at least one point.

II. **Axioms of distance**

II_1. For every two points A and B, there corresponds a nonnegative magnitude $|AB|$, which is called the distance from point A to point B.

$$|AB| = 0 \quad \text{if and only if } A = B.$$

II_2. The distance from point A to point B is equal to the distance from B to A:

$$|AB| = |BA|.$$

II_3. For any points A, B, and C

$$|AC| \leq |AB| + |BC| \text{ (triangle inequality).}$$

III. **Axioms of order**

III_1. Any point O on a straight line p divides the set of points on the straight line that are not O into two non-empty subsets of points in such a way that (a) for any two points A and B that belong to different subsets, point O lies between A and B; (b) for any two points A and B that belong to the same subset, one of them lies between the other point and O.

III_2. For any distance a on a given ray with its origin in O, there exists one and only one point whose distance from O is a: $|OA| = a$.

III_3. If a point C lies between points A and B, then the points A, B, and C belong to one straight line.

III_4. Any straight line p divides the set of points in the plane that do not belong to it into two non-empty sets in such a way that (a) any two points that belong to different sets are separated by the line *p*; (b) any two points that belong to the same set are not separated by the line *p*.

IV. **The axiom of congruence**

For any two pairs of points A and B and A′ and B′, such that |AB| = |A′B′|, there exists two "rigid motions" that take A into A′ and B into B′.

V. **The axiom of parallelism**

Through any point À in a plane, there passes not more than one straight line that is parallel to a given straight line.

Remark. All of the concepts used in the formulations of the axioms are subsequently defined in the text. Point X lies between two different points A and B if |AB|+|XB| = |AB|. A segment is defined as the figure consisting of two points and all points lying between them. "Rigid motion" is the mapping of a plane onto itself that preserves distance, etc. Kolmogorov laid out the axiomatics of scalar quantities in his article "Quantity" in the *Great Soviet Encyclopedia*.

Kolmogorov's axiomatic system made it possible to implement his conception of the course. The axioms of distance are the axioms of a metric space. The axiom of congruence makes it possible to begin talking about Felix Klein's ideas. The formulation of the axiom of parallels is the traditional approach to grasping the ideas of Lobachevsky.

It is important to emphasize that distance is not a number, but a magnitude. Such a view was accepted in Euclid's time. Kolmogorov attributed a great deal of meaning to the concept of magnitude. He developed an axiomatics of scalar magnitudes, and in a course for the Physics–Mathematics School he used this as a basis on which to construct a theory of real numbers, in which positive numbers are defined as monotonic additive operators on the set of scalar magnitudes (Kolmogorov, 1966b). None of this was mentioned in the new school curriculum, of course: the textbook merely called attention to the fact that numerical values of magnitudes depend on the choice of the unit of measurement. The discussion of magnitudes in the course on geometry paved the way for a discussion of dimensionality in the course on physics.

In 1970, when I became Kolmogorov's graduate student, he proposed that I take up the topic of "The Logical Foundations and Plan of Geometry." The aim was to work on the foundations of geometry, i.e., to construct a sequence of definitions and proofs for the main theorems of Euclidean plane geometry. He was interested above all in the theorem on angle measurement, which had remained without a proof. This work was completed in 1975, when I defended my thesis, which also contained many comments on the school curriculum. The material from this dissertation became the basis for specialized courses at pedagogical institutes.

An optional chapter entitled "The Logical Construction of Geometry" appeared in an edition of the eighth-grade textbook (1974–1977). Here, Kolmogorov offered a brief and accessibly written overview of the axiomatic method, including a discussion of the consistency, completeness, and independence of axioms, illustrated with examples from finite geometry. At the Physics–Mathematics School in the early 1970s, he taught a wonderful course in which geometry was constructed on the basis of axioms of incidence. The course began with an examination of finite affine and projective planes and spaces. An outline of Kolmogorov's lectures has survived, but has not been published.

The edition of 1972–1974 did not solve the problems that arose when the textbook was used in schools. Teachers absorbed the course slowly and with difficulty. During these years, there appeared a new genre of published literature that could offer practical help to teachers: lesson-by-lesson methodological analyses which contained very detailed recommendations on how to organize every lesson.

Work on a new version of the textbook "Geometry 6–8," which began in 1973, lasted much longer than expected. The unification of three textbooks in one book precluded the possibility of transferring problems from one year to the subsequent year. Several transitional drafts were prepared. The authors systematically searched for a more compact structure, and simpler methods and proofs. The system of problems went through substantial revisions.

Finally, a unified textbook came out in 1979 (Kolmogorov *et al.*, 1979). But this textbook did not last long — its last edition came out in 1982, at which point a political decision was made to replace it with a textbook by A. V. Pogorelov (see the section on the "counter-reform" below).

I actively participated in the preparation of the 1979 edition as a co-author. The decision to replace this textbook probably was premature. Over 10 years of work, teachers had accumulated considerable experience; the situation had begun to improve. In the early 1980s, Kolmogorov and I actively discussed ways to improve the textbook; a new prospectus was prepared. But these plans were never realized. A. F. Semyonovich and R. S. Cherkasov, who continued working on the textbook, submitted their new version to a competition in 1987, but their project did not meet with success.

I will now turn to the space geometry textbooks. Two groups of authors, which had emerged from a competition held in 1964, competed with one another during the developmental stages. The first group included teachers from the Kursk Pedagogical Institute, V. M. Klopsky and M. I. Yagodovsky. The author of the second textbook was K. S. Barybin, a methodologist from Moscow. At the beginning of the reform, the first group was enlarged to include Z. A. Skopets, a professor at the Yaroslavl Pedagogical Institute. Skopets, who was the author of famous problem books in geometry, became the textbooks' co-author and scientific editor (Klopsky, Skopets, and Yagodovsky, 1969, 1971). The second textbook was edited by A. B. Sossinsky, an associate professor at the Moscow State University mathematics department (Barybin, 1970, 1971).

In the relatively unanimous opinion of numerous reviewers, Barybin's textbook, which contained a multitude of mathematical inaccuracies, failed to reflect the ideas of the reform. Experiments with putting this textbook into use pointed to the same conclusion. Consequently, it was rejected.

In the mid-1970s, I worked at the publishing house "Prosveschenie" and edited the first, large-scale editions of the textbook by Z. A. Skopets *et al.* (Klopsky, Skopets, and Yagodovsky, 1975, 1976). I must say that this group of authors was made up of very experienced, highly qualified, and conscientious people who produced a sound textbook.

In terms of its contents, the textbook adhered to the principles described above — the principles employed in the course in plane geometry. The foundation for the course in space geometry consisted of Kolmogorov's axiomatics, supplemented by spatial axioms of incidence. Definitions were kept rigorous and formulations precise, although adhering to this approach when dealing with complicated notions such as "vector," "polyhedron," and "volume" made things very difficult for the authors, and subsequently for the students and teachers as well. Students were systematically taught to conceive geometrical figures as sets of points.

New methods, which had been prefigured in classes 6–8, were actively developed both in theory and in problem solving. Considerable attention was devoted to vector methods for solving problems (these methods were explicitly emphasized); the notion of scalar product appeared as well. But the idea of constructing the course in geometry on a purely vectorial foundation — which was fashionable at the time — was not even discussed.

The isometry classification theorem for three-dimensional space was not formulated, but certain types of isometries of space were discussed in theory and applied in solving problems. Problems "on visualizing symmetry," which were included in the course, were employed specifically in order to develop spatial imagination. Nonetheless, problems such as "How many axes (planes) of symmetry does a cube have?" invariably created difficulties for students and teachers alike.

By comparison with the traditional approach, the new textbook greatly simplified the derivation of formulas of volume by using the notion of integral. The area of the sphere was determined using Minkowski's method: as the derivative with respect to the radius of the volume of the ball that it bounds.

Sets of problems on constructing the cross-sections of polyhedra and problems on "imaginary constructions" helped to develop spatial imagination. For the first time, rules for representing spatial figures on a plane appeared in ordinary schools: the properties of parallel projection were formulated and proved.

The beginning of the course followed a more or less traditional approach: theorems about the mutual positions of straight lines and planes in space were proved on the basis of axioms of incidence; and related problems were discussed, with the aim of developing students' deductive skills and spatial imagination.

On the whole, it must be noted that, despite the introduction of many new topics, the course in space geometry drew less criticism from teachers than the course in plane geometry. This is explained, first, by the fact that many of the ideas developed in the upper grades had already been explained in grades 6 and 7 in a preliminary fashion. Second, with respect to methodology, the textbook was quite good, and a system of problems had been worked out that contained a "spectrum of assignments" from simple to difficult ones. Methods for solving problems were formulated clearly and precisely, and they were accompanied by examples. In addition, a substantial body of problems that were familiar to teachers from A. P. Kiselev's old textbook were retained. Nonetheless, difficulties did arise. They were connected with the conceptual intensity of the course and the shortage of class time.

Despite its strengths, the textbook edited by Z. A. Skopets was quickly replaced during the "counter-reforms" of the 1980s. As in the case of Kolmogorov's textbook, I believe that rejecting it completely was a mistake. Despite initial difficulties, teachers were beginning to get used to the new curriculum. The last editions of the textbooks, substantially revised, were free of many of the defects present in earlier versions.

9 Algebra and Elementary Calculus

In keeping with the program of 1968, the modernization of the course in algebra initially was moderate in character. Conspicuously greater attention was paid to functions and graphs. Set theoretical concepts and symbols were used. The main innovation consisted in shifting the topic "Exponential Functions and Logarithms" and the advanced sections to the eighth grade. Thus, the overall contents of the course "Algebra 6–8" was well known to teachers. This meant that the new textbooks were accepted and put into use with little difficulty (Makarychev, Mindiuk, and Muravin, 1972, 1973, 1974).

One should also note that the group of authors who wrote these textbooks — by contrast with the groups of authors who worked on the other textbooks — did not include any professional mathematicians and consisted entirely of methodologists (N. G. Mindyuk, Yu. N. Makarychev, and S. B. Suvorova). The textbooks' editor, A. I. Markushevich, was, of course, a professional mathematician, but he had a very moderate notion of modernization. In addition, the authors were realistic about how difficult the problems in the textbook could be. Due to all of these circumstances, the new textbooks in algebra were favorably received by teachers.

Probably the only fundamental difficulty arose in connection with the fact that the authors — who were very diligent and conscientious methodologists — wished to explicate all concepts fully; as a result, they somewhat overloaded their textbooks with problems aimed at testing students' understanding of the concepts of "mapping" and "correspondence." Consequently, in the late 1970s, the algebra textbooks that had been edited by A. I. Markushevich were subjected to significant criticism. At the beginning of the 1980s, when S. A. Telyakovsky (a researcher at the Mathematics Institute of the USSR Academy of Sciences) became the new editor of the series, the shortcomings were eliminated in new editions of the textbooks (Makarychev, Mindiuk, and Suvorova, 1981, 1982, 1983). Following a competition in 1987, these textbooks were recommended for use in schools across the country. Beginning in the following year, they started to be used in schools along with the textbooks of Alimov, Kolyagin *et al.* (1988, 1989a,b, 1990), which had also received honorable mentions in the competition.

The earliest experience with teaching calculus in Russian schools dates back to the beginning of the 20th century. Elements of differential calculus were part of the curriculum in "real-schools"; A. P. Kiselev (1908) wrote a calculus textbook. Strangely, this historical episode exerted a substantial influence on decisions made in the 1980s. In discussions of the mathematics curriculum, many people favored omitting calculus from the program. The case was decided by the opinion of the academician I. M. Vinogradov, who himself had graduated from a "real-school": "We already studied calculus before the Revolution. So we need calculus in the schools."

Elementary calculus was introduced into schools across the country in 1965, when the Kochetkovs' textbook went into use (Kochetkova and Kochetkov, 1965). This textbook, which was written by famous methodologists, was rather sloppy from the mathematical point of view; therefore, Kolmogorov agreed to edit a text written by B. Ye. Veits and I. T. Demidov (1969, 1970), teachers at the Murmansk Pedagogical Institute. In terms of mathematics, this book was better than the other one. After a comparative study of the two textbooks, Veits and Demidov's text was recommended for use in schools.

At the same time, it was recognized that this text was rather difficult for ordinary schools, particularly the sections concerned directly with calculus. The authors effectively adhered to the method of exposition used in colleges (limits of series — limits of functions — continuity, etc.). This textbook was taken as a foundation, and the group of authors was substantially enlarged to include Kolmogorov (as author and editor), S. I. Shvartsburd, O. S. Ivashev-Musatov, and B. M. Ivlev. The authors produced a new version of the textbook for use in ordinary schools (Kolmogorov *et al.*, 1973a,b, 1974a,b). Although the text was simplified, it still gave rise to considerable difficulties in schools; therefore, work on it continued in 1978–1979, with the final edition appearing in 1980 (Kolmogorov *et al.*, 1980).

The 1980 edition contained the following key changes. (1) Due to a reduction of class hours, sections on combinatorics and the principle of mathematical induction had to be omitted. (2) The exposition of key concepts (limit, derivative, and integral) was greatly abridged and simplified. (3) Due to severe criticism of the textbooks by the USSR Academy of Sciences (see below), the use of set theoretical concepts was reduced.

The next revision (Kolmogorov *et al.*, 1988) was completed in 1987, when a textbook competition was organized. At this time, the following fundamental changes were made:

1. The basic concepts of calculus (continuity, derivative, and integral) were formulated on the basis of illustrative geometrical and physical notions; attempts to provide a precise definition of limits were abandoned.
2. The system of problems and the methodological apparatus were substantially developed: (a) questions for review and model problems were added to each chapter, thus specifying what exactly students were expected to know; (b) the sections containing "Historical Facts" were systematized; (c) while the new edition as a whole placed lower demands on students' knowledge, it was necessary to preserve texts that were aimed at students who were interested in mathematics — to this end, a set of "Advanced Problems" was compiled and published initially as a supplement to the textbook, and subsequently as part of the text.

This textbook, edited by Kolmogorov and still in use, has not been revised in any significant way since 1987.

Intensive work on developing alternative textbooks began in the early 1980s. One of them was prepared as part of a large project headed by the academician A. N. Tikhonov. An experimental textbook by a new group of authors began to be published in the late 1970s (see, for example, Alimov *et al.*, 1984). Its distinctive features were a minimization of facts about calculus and a search for the simplest methods of exposition.

Another textbook was produced by M. I. Bashmakov, the famous Leningrad mathematician who worked on the problems of school-level education. This textbook was characterized by its conciseness and attention to practical application (Bashmakov, 1989).

Following the competition of 1987, all three textbooks were recommended by the Ministry of Education. The Kolmogorov textbook and the Bashmakov textbook won second place, while the textbook by Alimov *et al.* won third place (first place was not awarded). The results of the competition were published in the journal *Matematika v shkole* (1987, no. 1; 1988, no. 2).

10 The Counter-Reform

The events of the late 1970s and 1980s may be characterized as a counter-reform, since their logic was determined by sharp criticism of the still-incomplete reform of mathematics education and a wish to revise the transformations that had already taken place.

A decisive role during this period was played by the mathematics division of the USSR Academy of Sciences; the counter-reform was spearheaded by academicians I. M. Vinogradov (director of the Steklov Mathematics Institute), L. S. Pontryagin, A. N. Tikhonov (director of the Institute of Applied Mathematics). Teachers and methodologists who had invested a great deal of labor in the development of the "new school mathematics" were beginning to get used to the new textbooks and were not opposed to the Kolmogorov reform, but several groups that could potentially support the counter-reform movement were roused into action in the course of a discussion that had been initiated by mathematicians.

I believe that it is not an accident that the counter-reform began specifically in 1978. One year earlier, high-school students who had been educated in schools that implemented the reforms started taking college entrance exams for the first time. The colleges were confronted by a serious and acute problem: how to organize the exams?

There were two alternatives. The first was to preserve the already well-developed style of the exams, which often consisted of artificial problems that required students to use extremely intricate technical methods for solving them. It should be noted that this approach led many colleges to develop their own, idiosyncratic traditions, which greatly increased demand for the services of tutors and college instructors. This phenomenon became so prominent that a new name was invented for the "science" of writing problems for college entrance exams and tutoring methodology — "college-ology" (*"vuzomatika"*).

The second option was to undertake the difficult and serious work of substantially modernizing the system of college entrance exams in light of the radical changes that had taken place in the schools, and of making the transition to a system of problems that tested students' knowledge of mathematics and inventiveness, their readiness for college-level studies, rather than their ability to pass a specific college's entrance exam. But this option, which was supported by Kolmogorov, was much more difficult than the first, which determined many college workers' attitudes toward the reform.

The most influential group that came to support the counter-reform movement was the RSFSR's Ministry of Education. When the USSR Ministry of Education was established, tensions quickly arose between the education ministries of the USSR and the RSFSR, as often happens when two bureaucratic organizations fight over a sphere of influence. (There was even a joke in educators' circles: it is a good thing that there is only a USSR Ministry of Defense and no Ministry of Defense for the RSFSR. The joke turned out to be prophetic: in 1990, the RSFSR Ministry of Defense was created, and the collapse of the Soviet Union soon followed.)

The discussion about mathematics in the schools allowed RSFSR education minister A. I. Danilov (later replaced by N. V. Aleksandrov and G. P. Veselov) to assume an independent and aggressive position.

As shown above, there were plenty of grounds for constructive criticism. But the harshness with which the Academy of Sciences' mathematics division came out against the reforms was hardly justified. In 1967, the mathematics division had approved the plan for the new curriculum, and the basic goals of the school reform had been supported at a joint meeting of the presidiums of the Academy of Sciences and the Academy of Pedagogical Sciences (1970), chaired by M. V. Keldysh, the president of the Academy of Sciences. During the 1970s, academic circles took no part in the reform and were not interested in it.

Undoubtedly, subjective factors exerted a strong influence on the way in which events unfolded. By this time, for a number of reasons, Kolmogorov's personal relations with I. M. Vinogradov, L. S. Pontryagin, and A. N. Tikhonov had become quite complicated. There were also certain differences in their views of mathematics. Kolmogorov's diary contains the following entry, dating back to 8 January 1944: "Graduate student committee with Pontryagin and Plesner. Total chaos. Pontryagin keeps picking on Fomin and

Millionshchikov (not without justification, but with a special antipathy toward set-theorism)."

The counter-reform was launched in May 1978, during a discussion of school-related problems at a special meeting of the board of the Academy of Sciences' mathematics division, chaired by M. V. Keldysh. (This was practically the only time that Keldysh made a speech criticizing the reforms. He died in June of that year.) Kolmogorov was invited to attend the meeting. A critical resolution was passed and the decision was made to hold a special meeting of the mathematics division wholly devoted to school-related issues. This meeting took place on 5 December 1978. I was present at it, accompanying Kolmogorov, and have described the event in detail in a pamphlet "On the Situation of Mathematics Education in Secondary Schools in the USSR" (Abramov, 2003).

In preparing for the meeting, Kolmogorov made no attempt to draw his colleagues' votes over to his side. He spent a long time preparing his presentation and he prepared for it thoroughly — uncharacteristically for him, he prepared not just the key points, but a complete text (see Abramov, 2003). In his speech, he analyzed the situation quite critically and gave his assessment of the textbooks (criticizing in particular the textbook on space geometry). He indicated weak points and outlined a program of action.

The other speeches and presentations were quite critical. The harshest remarks were made by L. S. Pontryagin and A. N. Tikhonov. Only L. V. Kantorovich and S. L. Sobolev came out in favor of the ideas of the reform, calling for moderation. The final resolution, which was passed virtually unanimously, was quite severe: "The existing state of curricula and textbooks is acknowledged to be unsatisfactory." A committee on mathematics education — to be chaired by I. M. Vinogradov — was formed, and support was voiced for the RSFSR Ministry of Education's idea to develop a new curriculum and new textbooks, and to begin testing them out in practice (Abramov, 2003).

Subsequent decisions (1982) were made by Central Committee of the CPSU: geometry textbooks edited by Kolmogorov and Z. A. Skopets were removed from schools; A. V. Pogorelov's textbook was introduced in an accelerated fashion. These decisions were undoubtedly influenced by I. M. Vinogradov and A. N. Tikhonov, who were in direct contact with very powerful people in the Central Committee.

A considerable role both in shaping public opinion and in acquiring support for the "counter-reformers" in top government circles was played by L. S. Pontryagin. In 1979, he published an article entitled "Ethics and Arithmetic" (*Sotsialisticheskaya industriya*, 26 May 1979), in which, without actually referring to Kolmogorov by name, he accused him of irresponsibility and immorality. The article provoked a significant response, but no organizational steps were taken as a result of it. What tipped the scales in favor of the critics of the reform was an article that L. S. Pontryagin published in the main ideological journal of the Central Committee of the CPSU, *Kommunist* (1980, *14*). Through this article, the issue was elevated to the realm of ideology (although Kolmogorov's name was, again, not mentioned) and consequently now required a decision by the top leadership of the country. The political significance of the problem of school mathematics was emphasized by the academician A. A. Logunov, President (rector) of Moscow State University, in a speech at a session of the Supreme Soviet of the USSR.

The mood which determined the actions of the "counter-reformers" is well illustrated by an episode at which I was present. Somewhat unexpectedly, in May 1980, I found myself together with V. V. Firsov in the office of I. M. Vinogradov, who had by then already turned 90. As soon as the meeting began, Vinogradov summoned practically all of the people present in the institute — very famous members of the Academy. Vinogradov sat at the head of the table; V. V. Firsov and I, as guests, sat across from L. S. Pontryagin.

Vinogradov posed a question with which he was preoccupied: "Could we not replace these anti-government textbooks by September 1?" (i.e., within three months). Pontryagin said that unfortunately this was impossible — three months was not enough time to publish many millions of new textbooks, which would also have to be written first.

The difficulty of the "counter-reformers"' task resided in the lack of an alternative. M. A. Prokofiev, the USSR education minister, understood perfectly well that sudden changes were inadmissible in the inevitably conservative, gigantic system of secondary education (in the USSR, there were over 40 million schoolchildren), and diplomatically, but quite concertedly, opposed attempts at radical and rapid changes.

The opposition understood the need for an alternative as well. In 1979, two promptly prepared drafts for a mathematics curriculum were published. One of them was created by I. M Vinogradov's committee (*Matematika v shkole*, 1979, *2*), the other by a committee at the RSFSR Ministry of Education, headed by A. N. Tikhonov (*Matematika v shkole*, 1979, *3*).

The only educational texts that could in principle be put to use in schools were two books on elementary geometry ("Plane Geometry" and "Stereometry") by A. V. Pogorelov (1969, 1970), which had been published during the 1970s by Fizmatgiz, the State Physics–Mathematics Publishing House. Correspondence between Kolmogorov and the academician A. V. Pogorelov, in which Kolmogorov reviewed these texts, has survived. Kolmogorov had a favorable attitude toward the books and had recommended them for publication.

The problem was that Pogorelov had written his books as textbooks for pedagogical institutes. This meant that the text would have to be adapted urgently for schools and that a system of exercises would have to be developed. This work took up about a year and in September 1980, the new textbooks (Pogorelov, 1980) were introduced on a trial basis in the cities of Sevastopol and Kharkov (where Pogorelov was working at the time), under the aegis of the education ministry of the Ukraine.

The "anti-Kolmogorov" coalition broke up relatively quickly. A. N. Tikhonov became the head of authors' groups created at the RSFSR Ministry of Education, thus acquiring great administrative power. L. S. Atanasyan (a professor at the Lenin Pedagogical Institute in Moscow and the author of geometry textbooks for pedagogical institutes) and three physics professors from Moscow State University — E. G. Poznyak, V. I. Butuzov, B. V. Kadomtsev — began writing new geometry textbooks in 1979 (see, for example, Atanasyan *et al.*, 1979). Textbooks in algebra for grades 6–8 and algebra and elementary calculus for grades 9–10 were written by Professor Sh. A. Alimov (A. N. Tikhonov's student); the well-known methodologist

Yu. M. Kolyagin, who would soon become a member of the Academy of Pedagogical Sciences; Moscow Institute of Physics and Technology professor M. I. Shabunin; and V. A. Ilyin, a physics professor at Moscow State University (currently a member of the Russian Academy of Sciences).

Vinogradov's committee (after his death in 1982, L. S. Pontryagin took over as chairman) supported other authors' groups. The decision was made to retain but substantially revise the textbook in algebra for grades 6–8. S. A. Telyakovsky, who at the time was the secretary of the "Vinogradov committee," became the science editor of the project. The academician S. M. Nikolsky became the head of an authors' group that produced a textbook in arithmetic for grades 5 and 6, and the textbooks "Algebra 6–8" and "Algebra and Elementary Calculus 9–10" (see, for example, Nikolsky *et al.*, 1984). M. K. Potapov, a mathematics professor at Moscow State University, and N. N. Reshetnikov, a researcher at the Academy of Pedagogical Sciences, joined this collective.

The academician A. D. Aleksandrov, who replaced Kolmogorov in 1980 as head of the Scientific Methodological Council, became the chairman of an authors' group that wrote textbooks in geometry. Professor A. A. Werner of the Leningrad Pedagogical Institute and V. I. Ryzhik, one of Leningrad's best teachers, became his co-authors (see, for example, Alexandrov, Werner, and Ryzhik, 1984).

Also in Leningrad, an authors' group chaired by D. K. Faddeev — an associate (corresponding) member of the Academy of Sciences — began working on an algebra textbook (see, for example, Faddeev, 1983). Preliminary materials were prepared, but the work was soon interrupted by Faddeev's illness.

11 The 1980s

The extraordinary activity of the scientists from the USSR Academy of Scientists created a situation that was fundamentally new for Soviet schools. The existing programs (Kolmogorov's, Vinogradov's, and Tikhonov's) were all different from one another. Even more importantly, the USSR had a tradition of using the same textbooks for the whole country. A new problem arose: How could textbooks be diversified? This question was answered in 1981 when a new mathematics curriculum was created (*Matematika v shkole*, 1982, *2*).

In 1980, V. V. Firsov was appointed director of the mathematics education laboratory at the USSR Academy of Pedagogical Sciences. He went on to exert a great influence on mathematics education in Soviet schools during the 1980s. In effect, the waning of Kolmogorov's influence and the creation of competing authors' groups headed by famous mathematicians resulted in a certain ideological vacuum. Under these circumstances, V. V. Firsov unexpectedly became a leading figure and in large measure determined the subsequent course of events. This was facilitated by his good mathematical education (Moscow State University mathematics department), his deep interest in school-related problems and his experience with working in Moscow State University mathematics circles, his intellectual freedom, and his communication skills. He was convinced that the only way out of the existing situation was through constructive action. To this end, it was first and foremost necessary to create good working conditions for many different authors' groups.

Firsov pinned his hopes on the Academy of Sciences' mathematics division and the USSR Ministry of Education. He spearheaded and actively participated in the preparation of A. V. Pogorelov's school-level textbook, repeatedly meeting with Pogorelov and convincing him that it was necessary to make substantial revisions (something that was not easy to do, since Pogorelov was difficult to convince). Firsov also spearheaded the writing of methodological supplements for Pogorelov's textbook. He had it tested out in practice. Firsov had no illusions about Pogorelov's text, but he believed that putting it into use was the lesser of all evils, considering the instability of the situation and the inferior quality of the other available textbooks. Firsov also supported an authors' group working on the textbook "Algebra 6–8" (edited by S. A. Telyakovsky). The USSR Ministry of Education instructed Firsov's laboratory to conduct a comparison between the knowledge levels of students who had been educated using different textbooks.

The development of the 1981 curriculum was probably the decisive event that stabilized the situation. This curriculum, which was developed by Firsov, N. N. Reshetnikov, and myself, was founded on the following ideas:

1. A comparison of the three existing plans revealed that, despite differences between the orders in which topics were arranged, disagreements about the role of set theory in a school-level course, and different notions about how much time to spend on each topic, they had a great deal in common. The body of knowledge that all three plans proposed to cover was largely the same, overlapping by roughly 90%. This revealed the possibility of a compromise and at the same time showed that — despite all the loud rhetoric — the academic community effectively had supported the basic principles of the 1968 curriculum.

2. In order for schools to be able to use different textbooks — given the fact that these textbooks presented topics in different sequences and devoted different amounts of time to the same topics — the following solution was proposed.

(a) The freedom of action allowed to authors' groups was restricted in the following way: For each stage of education (grades 1–3, 4 and 5, 6–8, 9 and 10), a universal, mandatory level of knowledge was established and the general requirements that students had to fulfill in order to pass from one stage to the next were explicitly formulated. All of this was specified in a section of the curriculum entitled "The Contents of Education."

(b) A sensible structure for the exposition of the material had to be found and headings had to be established for the sections into which the fundamentally new curriculum was to be subdivided. The decision was made to structure the curriculum along the principal substantive-methodological lines of the course in mathematics. Sections such as "Geometric Figures," "Elements of Calculus," etc., appeared, and the key concepts were distributed among them.

(c) Every set of textbooks was accompanied by a special (variable) section of the curriculum entitled "Subject Planning." This section constituted a curriculum in the familiar sense of the word, i.e., it described the methodology of exposition recommended for the various different textbooks, apportioned the material among different classes, and precisely scheduled the presentation of the subjects in each class throughout the school year.

In this way, the curriculum of 1981, for the first time in the history of Soviet schools since the 1930s, made it possible to teach using different textbooks. The section on "The Contents of Education" ensured the uniformity of education; the section on "Subject Planning" provided for its variety.

12 Epilogue

During the 1970s–1980s, the reform of mathematics education gave rise to many heated discussions; echoes of these arguments can be heard to this day. Without claiming to know the truth, I would like to express my view of the "pluses" and "minuses" of the reform.

1. Over the course of the 20th century, beginning with the work of the International Committee chaired by Felix Klein (1908), the mathematics community actively discussed various ideas of modernizing the contents of the school course in mathematics — introducing the elements of calculus, analytic geometry, and vector algebra. These ideas were first fully implemented in Russia during the "Kolmogorov reform."
2. During the reform of the 1960s–1980s, the literature for students and teachers became substantially richer and more diverse. The reform greatly stimulated the development of mathematics education methodology: many new ideas and names appeared. Much of what was done during those years remains relevant both for Russia and for other countries.
3. For students interested in mathematics, the reform was a positive phenomenon; however, its aim — to construct a conceptually rich and at the same time universally accessible school course in mathematics — was not achieved. The knowledge and skill levels of a considerable part of the students were lower than expected.
4. The relative failure of the reform had several causes.

First, mistakes of a substantial nature were made. One of the biggest among them was the sharp reduction of problems in arithmetic, which play a considerable role in the mathematical formation of schoolchildren and in their preparation for the study of algebra and geometry. Second, the very radical and rapid changes in the course in geometry were also a mistake. The set of problems and style of exposition were substantially altered — and many teachers turned out to be unprepared for these changes.

There is a widespread view that the reform was harmed by the elements of set theory that were introduced into the curriculum. I do not believe that this view is correct. There were, indeed, certain excesses, but they did not take up too much time and did not play a decisive role. The complete "eradication" of sets from the curriculum was an excessive over-reaction.

5. Many difficulties and negative results stemmed from the unrealistic nature of the objectives and time constraints imposed on the reformers by top government officials. Worldwide experience confirms that education reform inevitably requires extended periods of time for developing new content and preparing teachers. The two or three years from the creation of the new textbooks to their large-scale implementation were clearly not sufficient.

6. The situation was severely exacerbated by the decision to make universal education mandatory (1973). The decision to make the presentation of mathematical materials more scientific (1966) and the sudden, rapid growth in the scale of education were in conflict with one another.

7. A negative role was also played by the — mildly speaking — reserved attitude of the professional community toward the reform. Institutions of higher learning failed to restructure their systems of entrance exams, as envisioned at the beginning of the reform: the conceptual intensity of the new curriculum presupposed an easing in the requirements for technical skill. Even more significant was the absence of any restructuring in the contents of education at pedagogical institutes in accordance with the aims and goals of the reform.

8. The crisis of the late 1970s and early 1980s ultimately played a positive role: premises were created for putting an end to identical schools, curricula, and textbooks. New authors' groups were formed.

9. Finally, it should be noted that for all the drama of the history of the reform, today, 40 years later, schools mainly use curricula and

many of the textbooks that were created during the 1960s–1980s. On the other hand, it is also true that today this fact serves to hold back development. By the beginning of the 21st century, very serious events had taken place in the USSR and Russia that now call for very serious changes in school-level education.

I will conclude with a quote from the academician A. P. Ershov. Starting in the 1960s, Ershov worked a great deal on the problems of teaching informatics in schools, and in 1988, when a decision was made to bring informatics into the educational system, he became the leader of the new reform. Naturally, therefore, he was vitally interested in the reform of mathematics education.

In 1988, Ershov moved from Novosibirsk to Moscow and, in May, he asked me to visit him. He was preparing a talk for the International Congress on Mathematics Education, which was to take place in Budapest in August. It subsequently emerged that Ershov was at this time terminally ill (he died in December of that year). Our conversation, during which he asked me at length about the details of the "Kolmogorov reform," lasted several hours. In the end, he asked me to leave him a selection of documents.

Speaking about the reform of the 1960s–1980s in Budapest, Ershov said:

> The general situation, of course, is not a return to what we had 20 years ago. A new generation of successful mathematicians has been brought up on Kolmogorov's reforms. These individuals play a dominant role in the finest expressions of our mathematical thought and practice. In addition, the teachers, for all the difficulties that they have gone through, have been introduced to a great number of fresh and innovative ideas and have thus risen to a new level of self-awareness. A. N. Kolmogorov's activities stirred the creative energy of his academician colleagues, as a result of which the mathematical literature on school-level mathematics has become much richer. The journal *Kvant* came into being, along with its wonderful collection of supplementary volumes.
>
> I believe that we cannot assess the meaning, role, and fate of the Kolmogorov reform while confining ourselves to its scientific–methodological content. Its fate cannot be separated from the fate

of education as a whole — of the country as a whole — during that decade which our media delicately refers to as "the period of stagnation."

I would put it this way: if the Kolmogorov reform as a movement turned out to be a failure, then its failure represents nothing more than the projection onto mathematics of a more global failure of another grand movement, which consisted in the transition to mandatory secondary education with the retention of all of the former rigidity, homogeneity, and authoritarianism in the content and methodology of school-level education...

Thinking about the dramatic fate of the Kolmogorov reform and its conceptual leader, I cannot avoid drawing a parallel to the fate of another brilliant contemporary of Andrey Nikolayevich Kolmogorov. I have in mind the writer and poet Boris Leonidovich Pasternak and his main work, *Doctor Zhivago.* The same degree of talent, high professionalism, and capacity for ordinary work. The same incompatibility with many aspects of quotidian reality. The same inseparable connection with culture and with nature. The same extreme jealousy and prejudice on the part of his colleagues. The same exalted sense of his uncompromising predestination for some universal human mission... (Tikhomirov, 1999).

References

Abramov, A. M. (1988). O pedagogicheskom nasledii A.N.Kolmogorova [On A. N. Kolmogorov's pedagogical legacy]. *Uspekhi matematicheskikh nauk*, *43*(6), 39–74.

Abramov, A. M. (2003). *O polozhenii s matematicheskim obrazovaniem* [*On the Situation of Mathematics Education*]. M.: Fazis.

Abramov, A. M. (Ed.). (2008). *Kikoin, Kolmogorov. FMSH MGU* [*Kikoin. Kolmogorov. The Physics and Mathematics School of Moscow State University*]. M.: Fazis.

Alexandrov, A. D., Werner, A. L., and Ryzhik, V. I. (1984). *Geometriya. Uchebnoe posobie dlia 9–10 klassov* [*Geometry 9–10 (Textbook)*]. M.: Prosveschenie.

Alexandrov, P. S., and Kolmogorov, A. N. (1940). *Algebra. Chast' I* [Algebra. Part I]. M.: Uchpedgiz.

Alexandrov, P. S., and Kolmogorov, A. N. (1941a). Svoistva neravenstv i ponyatie o priblizhennykh velichinakh [Properties of inequalities and the concept of approximate magnitudes]. *Matematika v shkole*, *2*, 1–12.

Alexandrov, P. S., and Kolmogorov, A. N. (1941b). Irratsionalnye chisla [Irrational numbers]. *Matematika v shkole*, *3*, 1–15.

Alimov, Sh. A., Kolyagin, Iu. M. *et al.* (1988). *Algebra 6 (uchebnoe posobie)* [*Algebra 6 (Textbook)*]. M.: Prosveschenie.

Alimov, Sh. A., Kolyagin, Iu. M. *et al.* (1989a). *Algebra 7 (uchebnoe posobie)* [*Algebra 7 (Textbook)*]. M.: Prosveschenie.
Alimov, Sh. A., Kolyagin, Iu. M. *et al.* (1989b). *Algebra i nachala analiza 9–10* [*Algebra and Elementary Calculus 9–10*]. M.: Prosveschenie.
Alimov, Sh. A., Kolyagin, Iu. M. *et al.* (1990). *Algebra 8 (uchebnoe posobie)* [*Algebra 8 (Textbook)*]. M.: Prosveschenie.
Atanasyan, L. S., Poznyak, E. G. *et al.* (1979). *Geometriya. Probnyi uchebnik dlia 6-go klassa.* [*Geometry 6 (Provisional Textbook)*]. M.: Prosveschenie.
Barsukov, A. N. (1956). *Algebra 6–8*. M.: Uchpedgiz.
Barybin, K. S. (1970). *Geometriya 9 klass (probnyi uchebnik)* [*Geometry for Grade 9 (Provisional Textbook)*]. M.: Prosveschenie.
Barybin, K. S. (1971). *Geometriya 10 klass (probnyi uchebnik)* [*Geometry for Grade 10 (Provisional Textbook)*]. M.: Prosveschenie.
Bashmakov, M. I. (1989). *Algebra i nachala analiza 9–10* [*Algebra and Elementary Calculus 9–10*]. M.: Prosveschenie.
Bokovnev, O. A., and Shvartsburd, S. I. (Eds.). (1978). *Fakultativnyi kurs «Izbrannye voprosy matematiki 9 klass»* [*Elective Class on "Selected Questions of Mathematics for Grade 9"*]. M.: Prosveschenie.
Boltyansky, V. G., and Yaglom, I. M. (1964). *Geometriya 9* [*Geometry 9–10*]. M.: Prosveschenie.
Dopolnitelnye glavy po kursu matematiki 9–10 klassov dlya fakultativnykh zanyatii [*Additional Chapters for Elective Classes in the Course in Mathematics for Grades 9–10*]. M.: Prosveschenie, 1970.
Doneddu, A. (1979). *Evklidova planimetriya* [*Euclidean Plane Geometry*]. M.: Nauka.
Faddeev, D. K. (1983). *Algebra 6–8. Materialy dlia oznakomleniya.* [*Algebra 6–8. Informational Materials*]. M.: Prosveschenie.
Firsov, V. V. (Ed.). (1980). *Fakultativnyi kurs (10 klass) «Izbrannye voprosy matematiki»* [*Elective Class on "Selected Questions of Mathematics" (Grade 10)*]. M.: Prosveschenie.
Gusev, V. A. *et al.* (1972). *Geometriya v VI klasse. Posobie dlya uchitelya* [*Geometry in Grade 6. Teacher's Manual*]. M.: Prosveschenie.
Kiselev, A. P. (1908). *Nachala differentsialnogo i integralnogo ischisleniya* [*Elementary Differential and Integral Calculus*]. Moscow.
Kiselev, A. P. (1960). *Geometriya, Chasti. 1,2* [*Geometry, Parts 1, 2*]. M.: Prosveschenie.
Klopsky, V. M., Skopets, Z. A., and Yagodovsky, M. I. (1969). *Geometriya 9 klass (probnyi uchebnik)* [*Geometry for Grade 9 (Provisional Textbook)*]. Edited by Z. A. Skopets. M.: Prosveschenie.
Klopsky, V. M., Skopets, Z. A., and Yagodovsky, M. I. (1971). *Geometriya 10 klass (probnyi uchebnik)* [*Geometry for Grade 10 (Provisional Textbook)*]. Edited by Z. A. Skopets. M.: Prosveschenie.
Klopsky, V. M., Skopets, Z. A., and Yagodovsky, M. I. (1975). *Geometriya 9 (uchebnoe posobie)* [*Geometry for Grade 9 (Textbook)*]. Edited by Z. A. Skopets. M.: Prosveschenie.
Klopsky, V. M., Skopets, Z. A., and Yagodovsky, M. I. (1976). *Geometriya 10 (uchebnoe posobie)* [*Geometry for Grade 10 (Textbook)*]. Edited by Z. A. Skopets. M.: Prosveschenie.

Kochetkova, E. S., and Kochetkov, E. S. (1965). *Algebra i elementarnye funktsii. Uchebnoe posobie dlya 9–10 klassov* [*Algebra and elementary functions. Textbook for grades 9–10*]. M.: Prosveschenie.

Kolmogorov, A. N. (1966a). Ob uchebnikakh na 1966/67 uchebnyi god [On the textbooks for the 1966–67 school year]. *Matematika v shkole, 4.*

Kolmogorov, A. N. (1966b). *Vvedenie v analiz* [*Introduction into Analysis*]. M.: MGU.

Kolmogorov, A. N. (1967a). Ob uchebnikakh na 1966/67uchebnyi god («Algebra i elementarnye funktsii» E. S. Kochetkova i S. Kochetkovoy) [On the textbooks for the 1966–67 school year. "Algebra and elementary functions" by E. S. Kochetkov and E. S. Kochetkova]. *Matematika v shkole, 1.*

Kolmogorov, A. N. (1967b). K izmeneniyam v tekste uchebnika "Algebra dlya 6–8 klassov" A.N. Barsukova [About changes in the content of the textbook "Algebra for grades 6–8" by A. N. Barsukov]. *Matematika v shkole, 6.*

Kolmogorov, A. N., and Semyonovich, A. F. (1970). O probnom uchebnike geometrii dlya VI klassa [On the provisional textbook in geometry for grade 6]. *Matematika v shkole, 4.*

Kolmogorov, A. N. *et al.* (1970). *Geometriya 6 (probnyi uchebnik)* [*Geometry 6 (Provisional Textbook)*]. M.: Prosveschenie.

Kolmogorov, A. N. *et al.* (1971). *Geometriya 7 (probnyi uchebnik)* [*Geometry 7 (Provisional Textbook)]*. M.: Prosveschenie

Kolmogorov, A. N. *et al.* (1972a). *Geometriya 8 (probnyi uchebnik)* [*Geometry 8 (Provisional Textbook)*]. M.: Prosveschenie

Kolmogorov, A. N. *et al.* (1972b). *Geometriya 6 (Uchebnoe posobie)* [*Geometry 6 (Textbook)*]. M.: Prosveschenie

Kolmogorov, A. N. *et al.* (1973a). *Geometriya 7 (Uchebnoe posobie)* [*Geometry 7 (Textbook)*]. M.: Prosveschenie

Kolmogorov, A. N., Ivlev, B. M. *et al.* (1973b). *Algebra i nachala analiza 9* [*Algebra and Elementary Calculus 9*]. Edited by A. N. Kolmogorov. M.: Prosveschenie.

Kolmogorov, A. N. *et al.* (1974a). *Geometriya 8 (Uchebnoe posobie)* [*Geometry 8 (Textbook)*]. M.: Prosveschenie

Kolmogorov, A. N., Ivlev, B. M. *et al.* (1974b). *Algebra i nachala analiza 10* [*Algebra and Elementary Calculus 10*]. Edited by A. N. Kolmogorov. M.: Prosveschenie.

Kolmogorov, A. N., and Shvartsburd, S. I. (1975). Metod matematicheskoy induktsii [The method of mathematical induction]. *Matematika v shkole, 1.*

Kolmogorov, A. N. *et al.* (1979). *Geometriya 6–8* [*Geometry 6–8*]. M.: Prosveschenie.

Kolmogorov, A. N., Abramov, A. M. *et al.* (1980). *Algebra i nachala analiza 9–10* [*Algebra and Elementary Calculus 9–10*]. Edited by A. N. Kolmogorov. M.: Prosveschenie.

Kolmogorov, A. N., Abramov, A. M. *et al.* (1988). *Algebra i nachala analiza 9–10* [*Algebra and Elementary Calculus 9–10*]. Edited by A. N. Kolmogorov. M.: Prosveschenie.

Kolmogorov, A. N. (1988). *Matematika — nauka i professiya* [*Mathematics: Science and Profession*]. M.: Nauka, bibliotechka "Kvant."

Kolmogorov, A. N. (2007). *Izbrannye trudy. t.4: Matematika i matematiki. Kniga I. O matematike* [*Selected Works, v. 4, "Mathematics and Mathematicians," Book 1, "On Mathematics"*]. M.: Nauka.
Kolyagin, Iu. M. (2001). *Russian Schools and Mathematics Education* [*Russkaya shkola i matematicheskoe obrasovanie*]. M: Prosveschenie.
Makarychev, Iu. N., Mindiuk, N. G., and Muravin, K. S. (1972). *Algebra 6 (uchebnoe posobie)* [*Algebra 6 (Textbook)*]. Edited by A. I. Markushevich. M.: Prosveschenie.
Makarychev, Iu. N., Mindiuk, N. G., and Muravin, K. S. (1973). *Algebra 7 (uchebnoe posobie)* [*Algebra 7 (Textbook)*]. Edited by A. I. Markushevich. M.: Prosveschenie.
Makarychev, Iu. N., Mindiuk, N. G., and Muravin, K. S. (1974). *Algebra 8 (uchebnoe posobie)* [*Algebra 8 (Textbook)*]. Edited by A. I. Markushevich. M.: Prosveschenie.
Makarychev, Iu. N., Mindiuk, N. G., and Suvorova, S. B. (1981). *Algebra 6 (uchebnoe posobie)* [*Algebra 6 (Textbook)*]. Edited by S. A. Telyakovsky. M.: Prosveschenie.
Makarychev, Iu. N., Mindiuk, N. G., and Suvorova, S. B. (1982). *Algebra 7 (uchebnoe posobie)* [*Algebra 7 (Textbook)*]. Edited by S. A. Telyakovsky. M. : Prosveschenie.
Makarychev, Iu. N., Mindiuk, N. G., and Suvorova, S. B. (1983). *Algebra 8 (uchebnoe posobie)* [*Algebra 8 (Textbook)*]. Edited by S. A. Telyakovsky. M.: Prosveschenie.
Markushevich, A. I., Maslova, G. G., and Cherkasov, R. S. (Eds.). (1978). *Na putyakh obnovleniya shkolnogo kursa matematiki* [*Toward a Revision of the School Course in Mathematics*]. M.: Prosveschenie.
Mathematics Curricula (Draft) [Programmy po matematike (proekt)]. (1966). M.: Prosveschenie.
Moise, E., and Downs, F. (1968). *Geometriya* [*Geometry*]. M.: Mir.
Moro, M. I., Bantova, M. A., and Beltiukova, G. V. (1968). *Matematika I* [*Mathematics I*]. M.: Prosveschenie.
Moro, M. I., Bantova, M. A., and Beltiukova, G. V. (1969). *Matematika II* [*Mathematics II*]. M.: Prosveschenie.
Moro, M. I., Bantova, M. A., and Beltiukova, G. V. (1970). *Matematika III* [*Mathematics III*]. M.: Prosveschenie.
Nikitin, I. N. (1956). *Geometriya 6–8* [*Geometry 6–8*]. M.: Uchpedgiz.
Nikolsky, S. M., Potapov, M. K., and Reshetnikov, N. N. (1984). *Algebra 6 (probnyi uchebnik)* [*Algebra 6 (Provisional Textbook)*]. M.: Nauka.
Nikolsky, S. M., Potapov, M. K., Reshetnikov, N. N., and Shevkin, A. V. (1988). *Arifmetika* [*Arithmetic*]. M.: Nauka.
Novoselov, S. I. (1956). *Trigonometriya dlya 9–10 klassov* [*Trigonometry for Grades 9–10*]. M.: Uchpedgiz.
Nurk, E. R., and Telgmaa, A. E. (1988). *Matematika 4–5. Uchebnik dlya sredney shkoly* [*Mathematics. A Textbook for the 4–5 grades*]. M.: Prosveschenie.
Pogorelov, A. V. (1969). *Planimetriya* [*Plane geometry*]. M: Fizmatgiz.
Pogorelov, A. V. (1970). *Stereometriya* [*Space geometry*]. M: Fizmatgiz
Pogorelov, A. V. (1980). *Geometriya 6–10* [*Geometry 6–10*]. M.: Prosveschenie.
Prokofiev, M. A. (1997). *Poslevoennaya shkola Rossii* [*Postwar Schools in Russia*]. M.: Miros.
Rybkin, N. A. (1960). *Sbornik zadach po geometrii* [*Collection of Problems in Geometry*], parts. 1 and 2. M.: Prosveschenie.

Shiryaev, A. (Ed.). (2003). *Kolmogorov (k 100-letiiu A. N. Kolmogorova)* [*"Kolmogorov" (three-volume work on the hundredth anniversary of A. N. Kolmogorov's birth)*]. Moscow: Fizmatlit.

Shtokalo, I. Z. (1975). *Istoriya matematicheskogo obrazovaniya v SSSR* [*History of Mathematics Education in the USSR*]. Kiev: Naukova dumka.

Sikorsky, K. P. (1969). *Dopolnitelnye glavy po kursu matematiki 7–8 klassov dlya fakultativnykh zanyatii. Sbornik statey* [*Additional Chapters for Elective Classes in the Course in Mathematics for Grades 7–8. Collection of Articles*]. M.: Prosveschenie.

Tikhomirov, V. M. (Ed.) (1999). *Yavlenie chrezvychainoe (kniga o Kolmogorove)* [*An Extrarodinary Phenomenon (A Book about Kolmogorov)*]. M.: Miros-Fazis.

Veits, B. E., and Demidov, I. T. (1969). *Algebra i nachala analiza 9* [*Algebra and Elementary Calculus 9*]. Edited by A. N. Kolmogorov. M.: Prosveschenie.

Veits, B. E., and Demidov, I. T. (1970). *Algebra i nachala analiza 10* [*Algebra and Elementary Calculus 10*]. Edited by A. N. Kolmogorov. M.: Prosveschenie.

Vilenkin, N. Ya. *et al.* (1968). *Matematika 4* [*Mathematics 4*]. Edited by A. I. Markushevich. M.: Prosveschenie.

Vilenkin, N. Ya. *et al.* (1969). *Matematika 5* [*Mathematics 5*]. Edited by A. I. Markushevich. M.: Prosveschenie.

4

Challenges and Issues in Post-Soviet Mathematics Education

Mark Bashmakov

St. Petersburg University, Russia

1 Introduction

The year 1985 may be considered as the start of a new period in the history of mathematics education. In March of that year, Mikhail Gorbachev became the head of state, and the ensuing "perestroika" ushered in great changes in the lives of Soviet citizens. The weakening of the Communist party and a growing democratic movement, the fall of the Soviet empire and the emergence of new (more or less) independent states, the adoption of a new Constitution and legislature: these are just a few of the major milestones of that period.

The first decade of the new era was especially difficult. The shift toward a free market economy was attended by a dramatic decline in quality of life for the great majority of ordinary citizens, including educators. The economic recovery of the following years had a significant impact on the school system, which was finally able to access the funds necessary for its development. Not enough time has passed to give an objective account of the progress made in that time period. This article offers instead a brief overview of the developments in mathematics education beginning in 1985, tracing the larger trajectory of the school system as a whole, analyzing materials pertaining directly to mathematics education, and assessing the results and the emerging trends of these developments.

The following events falling within the given period have had the greatest impact on the school system:

1. The school reform, aimed at bringing the system in line with a new political strategy. Some of the objectives outlined in the reform were eradication of the government's monopoly on education; greater control over educational policy to be granted to regional authorities; implementation of the principles of "diverse education" and "humanistic approach to education" (Dneprov, 1994).

2. The 1988 Nationwide Congress of Educators, which adopted a new definition of general secondary education with a primary aim of "promoting intellectual, moral, emotional, and physical development of the individual, developing his or her creative potential to the greatest extent possible."

Naturally, this formulation was very much in line with the democratizing tendency that was starting to infiltrate all spheres of social life; however, the school system was not ready for its implementation, both morally (educators were not prepared to put the principles into practice) and economically (the new direction called for new methods of organizing the educational process).

3. The 1986–1988 nationwide textbook competition.

4. The emergence of a new legal framework: The Education Act was the first piece of legislature introduced by the first President of the Russian Federation, Boris Yeltsin. This was a great source of inspiration for teachers and administrators but, regrettably, the new act had relatively little effect and no immediate application. A great deal of effort was put into the new Education Law. Its effect, however, at least in its initial form, was largely negative: on the one hand the law admitted the possibility of a system of non-governmental educational facilities (though this prospect was instantly mired in countless bylaws imposing strict regulations on non-governmental schools); on the other hand it cut down the term of mandatory public education, exposing schools and educators to a great deal of confusion.

5. The restructuring of the Academy of Pedagogical Sciences and its transformation into the Russian Academy of Education (1992–1993).

2 Important Events in Mathematics Education of the Late 1980s

How do the events and facts bearing directly on mathematics education fit into the general trajectory of that time period?

2.1 *Emergence of a "Uniform Standard"*

The general mathematics curriculum adopted in 1986 was the outcome of a long-standing opposition to the Kolmogorov reform, making official the naturally evolved standard of mathematics education that had enjoyed a high reputation throughout the Soviet Union as well as abroad. This standard was maintained by means of mandatory education, scientifically substantiated courses of study required for all institutions offering secondary education, including a wide network of vocational schools, and long-standing ties with leading academic institutions. In the 1980s this educational standard was given a theoretical basis in the form of the "uniform standard" policy. Several leading experts participated in the development of the policy, including "task forces" from the nation's largest pedagogical research institution — the Institute on Educational Content and Methods — headed by the mathematician V. M. Monakhov. The theoretical formulation of the "uniform standard" was supplemented by workbooks illustrating expected performance levels (Monakhov, 1983).

The demands of the uniform standard of education were not easily reconciled with the preservation of accumulated traditions. Potentially adverse effects of the new standard were tempered, on the one hand, by the promise of high-level education "for all" and, on the other hand, by a certain amount of vagueness built into the standard.

2.2 *The Textbook Competition*

The open contest to produce new texts in mathematics took place as part of the general textbook competition of the late 1980s. Several academic institutions took part in reading the manuscripts (including the Institute for Mathematics at the Soviet Academy of Sciences), alongside various federal and regional educational agencies. Naturally, the bulk of the competing pool was made up of previously existing textbooks slightly revised to comply with contest rules. A small number of new textbooks (and new authors) did emerge, however. The competition took place at a time when the "unified standard" policy was at the height of its popularity, and it was assumed that a single "optimal" textbook would be chosen in each competing subject. The democratizing tendency with its demand for "free choice" already was in ascendance. Curiously, the winner of the competition for the elementary school textbook was announced before the option of "parallel" or alternative textbooks had been recognized officially. This was a new text, written by two relatively unknown educators, E. R. Nurk and A. E. Telgmaa (1988). The textbook was immediately picked up for mass distribution and aggressive implementation.

Winners in the "middle school" and "high school" categories were announced only six months later, but this time the jury named not one, but three winning textbooks in each subject. This shift permitted teachers to continue working with familiar texts and course structures, which varied significantly from textbook to textbook. The variations in structure were especially evident in geometry courses (traditionally in Russia geometry is taught separately from algebra in the last five years of school). Distinctive "approaches" to subject organization typically were referred to by the names of the authors of textbooks: e.g., the Pogorelov approach, the Alexandrov approach, the Atanasyan approach, the Kolmogorov approach, etc.

2.3 *New Objectives*

With respect to the traditionally defined objectives of mathematics education we can distinguish three main focal points. Each of them may be characterized by a specific formulation. The first formulation is attributed to the mathematician and naval engineer A. N. Krylov (1965, p. 607): "Mathematics is a means to an end, a tool no different than a mechanic's file and caliper or a carpenter's axe and saw" (p. 607). These words were uttered during a lecture delivered to an audience of engineering students in the 1920s. The direction they describe may be called pragmatic or *utilitarian*. The second statement belongs to the renowned mathematician Jean Dieudonné (1972), member of the Bourbaki group: "Fundamentally, mathematics has no utilitarian purpose, it is an intellectual pursuit" (p. 13). This somewhat paradoxical statement from a major mathematician places the emphasis on the *intellectual development* of the individual. Finally we come to the third formulation, which belongs to the contemporary Russian mathematician Nikolai Vavilov (2003, p. 8): "The fundamental aim of mathematics education is to inculcate in the student the principles of intellectual integrity" (p. 8). This position emphasizes the *pedagogical* value of teaching mathematics.

Naturally all pedagogical systems at any point in time comprise all three of the positions outlined; but their distribution and emphasis may vary significantly from system to system. It can be said quite frankly that for a long period of time the Soviet school system tended largely toward the pragmatic pole.

In the general mathematics curriculum, which remained in effect up through the 1980s, the objective of mathematics education was formulated thus:

> The primary aim of the study of mathematics in secondary school is the attainment of a firm, conscious grasp of mathematical knowledge and skills indispensable to every member of today's society in his daily interactions and professional activities, such as will permit the study of related disciplines and the pursuit of further education (quoted from Bashmakov, 1996).

In the new era other objectives were rising to the surface.

2.4 *The Dawn of the Computer Age*

The potential applications of the emerging technologies in the field of education already were becoming clear at the start of the new era, though few could foresee the full range of possibilities in information exchange and processing that appeared just 10–15 years later. The major trends of applying new technologies in mathematics education were also evolving at this time. These early formulations are still relevant today, and very few of them have been realized to their full potential.

Without entering into a discussion of matters of information access, processing and storage, or questions of management of education or communications technologies, I will note that by the early 1990s there had emerged two distinct approaches to the application of computer technologies. One proceeded from "hardware": i.e., people conversant with available technologies were offering them to educators; and the other proceeded from educators, aware of classroom challenges and looking to technology for possible solutions.

2.5 *What Are They Doing?*

Speaking of the changes in mathematics education that took place in the post-Soviet space in the years following *perestroika*, it would be fitting to touch on parallel educational practices abroad; these could not fail to draw the attention of Russian educators and, to a certain extent, influence the Russian school system.

2.5.1 *Differentiated education*

While the Soviet school system of the 1980s pursued a policy of uniformity, those of Western Europe long had put into effect the principles of differentiated education. This was readily apparent in the French system, as French textbooks, aimed at different types of schools, gradually made their way into Russia.

2.5.2 *Regional education; the development of "community education"*

Across several countries education had not been strictly centralized, becoming instead the prerogative of regional government. Germany is a classic example of this system: here regional authorities carry much of the responsibility for managing the educational process and results differ significantly from one locale to another.

A similar tendency may be seen in countries where various community-based institutions are given considerable power over the educational process (the UK and Switzerland being the leading exponents of this approach).

2.5.3 *National standard*

Most of the developed countries facing a decline in student performance have taken to setting national standards. At the same time the unification of the labor market in Europe has advanced a globalizing agenda that includes coordinated curricula, similar both in content and in performance assessment. Several major international projects have been undertaken with the aim of creating a uniform educational realm (e.g., the *Comnet* program or the more recent *Leonardo* program).

2.5.4 *Alternative schools*

The standardization of the school system has clearly reached a critical point across all the developed nations. The number of young men and women falling outside the general scheme has grown at a catastrophic rate. Different causes may be at play (ranging from rising deviations from the norm in psychological development to fluctuations in immigration policy), but the social implications of this increase and its threat to social stability have been more or less the same. This circumstance has given rise to alternative education programs.

A typical example is the "City-as-School" program developed in the US in the 1980s, which allowed tens of thousands of young men and women, who had dropped out of the traditional system, to complete their high-school education. The International Network of Productive Schools (INEPS) created in 1990 focuses on applying the principles of productive education to personal and professional self-determination in adolescents.

3 Digression: A Few Personal Remarks

In outlining the state of mathematics education in Russian schools as it had evolved toward the close of the 20th century, I have tried to be as objective as possible. Going forward, it will be difficult to refrain from indulging a personal viewpoint, therefore, it seems appropriate to give some account of my personal involvement with the issues in question.

My primary research interests and accomplishments belong to a fairly abstract field of contemporary mathematics; however, a deep interest in education had been instilled in me very early on and has manifested itself over the years in several ways: from participating in organizing the national mathematical Olympiads (where I have served as vice-chairman of the committee for 15 years); to participating in founding the popular science magazine *Kvant*, published since 1970; to establishing a famous boarding school at Leningrad University where I taught for 10 years; to designing curricula and writing textbooks for vocational schools (1975–1989).

All of these activities were pursued in parallel with my primary work as instructor and then professor of mathematics at Leningrad University. The beginnings of the new historical period coincided with my decision to take up education as my primary work and research. The textbook competition of 1986–1988 had played an important role in this shift, particularly the case of the textbook for algebra and elementary calculus. The text I submitted took the leading place in the competition, ahead of well-known and widely accepted texts (Bashmakov, 1991). Subsequently, I faced the daunting task of crafting supplementary instructional materials to accompany the textbook.

During *perestroika* I had taken part in the work of the first democratically elected Council of Leningrad (the same council that restored to the city its original name — St. Petersburg). As the chairman of the Committee on Public Education, I worked to dismantle the old bureaucratic system. Major changes had to be made throughout the system, all the while keeping the schools open and functional and preserving accumulated teaching traditions. I believe we were largely successful in our goals.

In the course of the Council's work it became evident that the utopian ideas of some of the new politicians were not always practical. A number of compromising but very useful proposals were rejected; others were never implemented. As a result, the system of education managed to regress quickly to what it was before.

In 1993, I was named a full member of the new Russian Pedagogical Academy (bypassing associate membership), which shifted my focus entirely to pedagogical issues that included not only specific topics in mathematics education, but extended also to the broader concepts of social pedagogy.

4 In the Beginning Was the Word...

The next phase (inaugurated in the early 1990s) began with a series of proclamations. No one could remain indifferent to the statements of then-president Yeltsin concerning self-determination of regions and republics: sovereignty and self-rule were there for the taking.

These proclamations held a series of implications for the school system and educational policy.

4.1 *Freedom to Design Curricula and School Programs*

In the early stages schools were left free to design their own curricula with minimal general requirements. For example, this was the time of so-called humanities gymnasia, which reduced mathematics credit requirements to a bare minimum. One such school, which attracted very talented and highly advanced students, had contacted our research group with a question: how can we use our 2 hours math requirement to provide adequate instruction to our students, most of whom had previously received high marks in mathematics and had retained an interest in and aptitude for mathematics? Naturally, such freedom was relatively short-lived, and a "tightening of the screws" followed soon after.

4.2 *Choice of Textbooks*

It was announced that teachers were free to choose texts from those offered for their courses. These offered texts, however, had to be approved by an expert committee at the Ministry of Education (at the federal level); and had to be in print. The first requirement was manageable, if not always especially efficient. The second requirement pre-supposed an expansion of the publishing sector. New publishing houses sprung up overnight. Publishers specializing in educational materials had to secure permissions from the authors of recommended textbooks and sign contracts with them. One important factor was that textbooks were paid for by the government, making such contracts highly profitable for the publisher. The system stumbled at the very outset: in order to get the green light from the Ministry of Education, a publisher had to comply with abundant technical requirements (e.g., health requirements dictating the choice of fonts, colors, layout, etc.) and win the contract of a regional authority that paid for the textbooks. All of this created an environment for profiteering, which seriously hurt the cause.

Overall, the previously existing system of printing and distribution of educational materials was dismantled, while the new one was mired in so-called "new economic methods" which proved to be largely ineffective.

4.3 *Methodological Support*

General methodological support of the school system had come traditionally from the Academy of Pedagogical Sciences and its research institutes. Rabid and unsubstantiated criticism of the Academy's work had been unleashed in the early 1990s, leading to the Academy's transfiguration into the Russian Academy of Education, which also absorbed all previously existing research institutes. These developments had little effect on the activities of the leading experts in education, who continued their research and experimental work.

Newly generated textbooks often were linked to a specific theoretical platform. As a result, most of the teachers became acquainted with competing ideas in contemporary pedagogy: e.g., incremental development of mental faculties (the Galperin–Talyzina system), problem education (the Makhmudov system), enlarged didactic units (the Erdniev system), developmental education (the systems of Zankov, Davydov, Elkonin, etc.).

At the same time, experimental systems proposed by "innovators," such as Shatalov, Lysenkova, and Schetinin (Fridman, 1987), elicited a great deal of support and interest among educators.

There were also major shifts in the field of education psychology. The powerful school of psychology that had predominated in Russia since the 1920s (it will suffice to recall the names of Vygotsky and Leontiev) continued its work, focusing during this period on placing its vast theoretical knowledge in the service of the educational system.

5 Word and Deed

The educational process is peculiar in that it cannot pause: schoolwork cannot be suspended while new ideas and principles of operation are prepared and implemented. Consequently, the transitional period began with the dismantling of the old Soviet school system, which had operated in a relatively stable socio-economic environment, through the end of "launch fever" was especially difficult for schools.

5.1 *Working Conditions of a Mathematics Teacher*

5.1.1 *The place of mathematics in school curricula*

Mathematics has always been the chief subject of study alongside native language and literature. It had risen in authority in the 1960s and maintained that position through the period under discussion. It then began to waver. Several external circumstances (in particular, economical), not to be discussed here, contributed to the shift. One less obvious factor that readily comes to mind is the changing psychology of the student. Up until a certain time the authority of the school and the teacher had been infallible. All of a sudden students could ask openly, "why do we need math at all?" And teachers could not find an answer that was convincing to the student or to themselves.

Simultaneously schools were reducing the required number of mathematics hours. In the 1970s high schools required a minimum of 420 hours of mathematics (about 6 hours per week) while the new curricula called for only 280 hours or even less for certain types of schools. Citing some concrete statistics: the 1952 general educational policy called for 64 week hours of mathematics education cumulatively across the student's entire school career. In 1985 the requirement had dropped to 60 hours; in 1998 it was merely 46 hours.

The precarious situation of mathematics was further aggravated by a general confusion over final examinations at all grade levels. The abolishment of oral examinations had a strong effect on classwork organization. Students were no longer required to speak, to solve problems at the board, and to discuss their solutions.

5.1.2 *The role of the teacher*

The national shift toward a market economy was accompanied at the outset by a *de facto* bankruptcy of the government and the draining of the national treasury, from which teachers' salaries previously had been paid. The new government pay scale made paupers of state employees. Anyone who could find work in the private sector did so, though there were many who could not. As a result, recent university graduates turned down teaching opportunities. A large number of educators, especially in metropolitan areas, left the profession, while the remaining teachers were forced to take on massive workloads in order to salvage their financial situation.

The situation gradually improved with the economic recovery. Many regions found additional funds to supplement their teachers' modest state salaries. Overall, however, the constant threat of poverty had gravely impacted the teachers' attitude toward their work. It was not possible to expect teachers to take the extra initiative to research or produce additional teaching materials, attend professional training courses, make use of outside information sources, such as libraries, research institutions, etc.

There were problems with textbook availability. Many of the regions could not afford to buy textbooks mandated by the central authority. Inevitably, families were asked to purchase their own textbooks, a situation made far worse by the skyrocketing prices of printed materials.

Methodological support weakened. Professional training specialists, likewise receiving meager state salaries, were more interested in giving paid lectures than in assisting working teachers in their day-to-day activities. Government agencies responsible for education management cut spending by gradually getting rid of educators on their staff.

These difficult circumstances had a certain positive effect as well: schools were becoming more independent as they struggled to survive. They became more sensitive to the specific needs of students and their families, whose role in a school's survival was becoming much more prominent.

5.2 *Educational Content*

5.2.1 *The structure of the educational process*

The study of mathematics in schools can be broken down into three phases:

Elementary school. In early education mathematics is integrated into the general course of studies taught by a single teacher. Generally mathematics takes up 4–5 classes per week.

Basic (middle) school. The five-year "basic school" period is divided into two cycles: grades 5 and 6, and grades 7–9. During the first cycle mathematics is taught as a single subject (4–5 classes per week). In grades 7–9 mathematics is divided into two distinct subjects: algebra and geometry, with an average of 5 classes per week devoted to both cumulatively.

High school. The high-school curriculum has been subject to numerous modifications since the introduction of differentiated education. The course is divided into two subjects: algebra with introduction to calculus, and geometry. The number of hours allotted to these subjects varied: typically schools required 5 classes per week, but some went much lower (as low as 3 hours per week) and others much higher (those schools that offered advanced mathematics training).

A crucial change to the curriculum, aggressively promoted by the central authority, was the differentiation of the curriculum into federal, regional, and local components. A great deal of effort was expended on the development of various elective courses, with the publication of lists of possible topics and course plans. Such courses frequently lay outside traditional curricula and teachers largely were unprepared to teach them. More importantly, elective courses did not address the problems that came with the reduction of time for mathematics classes (while overall course-load remained the same or even grew larger than before). As a result, many teachers were forced to use elective courses to complete material required by the main course.

5.2.2 *Minimum level*

In response to the judicious complaint that a traditional uniform mathematics curriculum cannot be covered in reduced time without significant loss in the quality of education, the school system introduced the concept of "minimum performance" levels. Problems in textbooks were marked with various multi-colored symbols, indicating skill levels: generally required, desired, suggested, etc.

In theory, this was supposed to help teachers, but in practice it only disoriented them. Worried about meeting minimum standards (which only could be defined vaguely), teachers started cutting back the time traditionally devoted to classwork components such as analysis, discussions, proofs, etc. Attempts to define minimum performance levels continue to this day, needlessly distracting schools from the true aim of seeking out new effective methods of instruction.

6 Modernization

As the nation was beginning to recover from the initial shock of the transition to new socio-economic conditions, both the society at large and the government paid increasingly more attention to problems of education. The term *modernization* was introduced at this time, signaling the introduction of certain new elements into the public education system. *Modernization* sounded less harsh than *reform*, a term used somewhat excessively in those days and evoking general antagonism to impending changes.

The term *modernization* covered a great deal of activity that found its way into schools and had relatively little effect there. Three elements of modernization, however, did have a strong effect on mathematics education and continue to exert their influence to this day. These may be characterized succinctly as *standards*, *specializations*, and *USE* (uniform state examination).

6.1 *Educational Standards*

The law "On Education" of 10 July 1992 dictates that

> primary and secondary school curricula aim at fulfilling the federal standard of education with respect to the type and specialization of the institution and the educational needs and demands of its students; they include general educational programs, outlines of courses, subjects, and disciplines (modules) along with other materials ensuring the moral development of students and quality of education (Legislative acts, 2008, pp. 16–18).

Since the passage of this law there has been an ongoing effort to give an adequate formulation of the standards of education. This problem remains unsolved to this day, and we will have a chance to discuss the prospects of its solution later on. For the moment we will limit ourselves to a formulation of the question.

According to the law "On Education" par. 4 (7) (including amendments introduced in 2007)

> Federal education standards dictate:
>
> 1) the structure of main educational programs, including the distribution of their components and their relation to the whole, as well as the distribution of required and elective components of programs;
> 2) the terms of administering educational programs, including matters of personnel, financial, logistical and other components;
> 3) assessment of student performance (Legislative acts, 2008, pp. 24–27).

The framers of the new version of the Standard (the so-called second-generation standard) explain the requirements as follows (Formulation of federal standards, 2008, pp. 22–23):

> *Requirements for structure of main educational programs*: this describes the totality of organizational and pedagogical conditions necessary for the realization of the educational process. It includes descriptions of general curricula and of their basic components, and dictates the overall scope of such curricula, the distribution of

their various components, and in particular the distribution of its mandatory and elective components.

Requirements for assessment of student performance: this describes the objectives of general education, formulated with a view of personal and societal needs and state goals. It describes the expected results of general education, determines its main directions and the specifics of its format and content.

Requirements for administering educational programs: this describes the totality of conditions (personnel, fiscal, logistical, prophylactic, etc.) necessary for the administration of general education.

The initial version of the Standard was relatively detailed. Criticism (pertaining to the mathematics component) was primarily leveled at the basic curriculum and subject content as outlined in the Standard. As has been noted before, the structure of the curriculum, including reductions in required mathematics clock hours, was thought to be the Standard's main weakness; however, much criticism was addressed to course content. The Standard dictated in considerable detail the minimum required content of the various courses by setting out the major themes, concepts, and facts.

Minimum requirements for graduates were formulated separately, for example, as follows:

At the completion of mathematics education at the basic level the student is expected to know/understand:

- — the significance of mathematical science in solving problems encountered in theory and in practice; the breadth and limitations of practical applications of mathematical methods to analysis and investigations of processes and phenomena in nature and society;
- — the significance of practical application and questions arising within the mathematical discipline itself for the development of the mathematical science; the history of the concept of number, the invention of calculus, and the emergence and developments of geometry;
- — the universal characteristics of the laws of logic in mathematical reasoning and their application in all spheres of human activity;
- — the probabilistic nature of various processes in the outside world.

Furthermore, each theme was followed by a section describing in what manner a student would be expected "to apply acquired knowledge and skills in practical life and everyday activities."

The advantage of the initial version of the Standard was a reasonably clear description of the objectives of mathematics education and a modest and not too strict list of expectations. Numerous complaints were directed at the list of themes, concepts, and facts, but these shortfalls of the Standard were not critical, though they did arouse a certain level of apprehension.

6.2 *Specialization of High Schools*

The concept of *class specialization* (or school specialization) was first introduced as schools were transitioning from a uniform standardized system of education to a differentiated one. Specialized mathematical schools, however, had been created much earlier under pressure from prominent physicists and mathematicians, whose opinions could not be ignored easily by the government. That early form of specialization did not affect general "non-specialized" curricula; it merely produced new curricula and textbooks designed for "advanced mathematics courses." Such courses often were based on college-level notions of mathematics education and the term "advanced" typically signaled the inclusion of college-level topics in a high-school course. In our opinion this innovation had a largely negative effect on the quality of instruction since it stymied radically new approaches to education by insisting on the adaptation and vulgarization of a college-level curriculum as the only possible means of modernizing high-school mathematics.

Specialization of high schools has become one of the main objectives of the ongoing modernization of the entire school system. The term *specialization* is typically interpreted as follows.

Any high-school student must select one specialized trajectory of education. Though there is a list of primary specializations, the number of specializations may be indefinite. Any subject may be studied at two levels: basic and specialized (advanced); i.e., specialization has become equated with the level of study. To make a specialized school (class), it is sufficient to decide which of the main subjects will be studied at the basic level and which at the advanced level. The advanced subjects will define the school (class) specialization.

Avoiding a general discussion of the effectiveness and the feasibility of such an approach, this chapter will focus on the problems pertaining to mathematics directly. Essentially, mathematics educators are asked to design a two-tier mathematics curriculum and schools are asked to choose either one tier or the other. In other words, this approach still maintains the idea that mathematical knowledge and skills may be arranged along a straight line and this line may be chopped down as necessary. Anything lying outside the straight line will be billed as "elective courses" and used in the manner of an optional condiment.

6.3 *Uniform State Examination (USE)*

State monitoring of the level and quality of education is a deep and complex problem and its solutions vary from state to state. Over a long stretch of time the Soviet Union had made use of a system of final examinations. The mathematics exam had two components: a centrally standardized written component and an oral component. Oral exams were made up of questions drawn randomly from a previously published list, with additional problems selected by the teacher. On the whole, in our opinion, this system functioned reasonably well. Problems appearing on the written exam were familiar to students, but more importantly, the grading of the exam lay largely in the hands of the school, while test preparation easily could be integrated into the rhythm of class work.

With the collapse of the Soviet Union and the ensuing process of decentralization, the school system was faced with the challenge of preserving a uniform educational environment. One of the solutions proposed was the USE. It would be hard to name another topic in education that provoked the same level of debate, both in professional and in lay circles. The mathematics component of USE went through a series of incarnations, but its great bulk was always made up of multiple-choice questions with five possible answers.

Art. 15 of the Law of the Russian Federation on Education (Legislative acts 2008, pp. 24–27) prescribes USE as a mandatory component of state certification of graduating students. Its aim is to provide "an objective performance assessment for individuals said to have completed a course of study" at different levels.

Just as a physical system is altered through measurement, so too the educational system is highly sensitive to the manner in which its efficacy is evaluated. No matter what form the assessment process takes, it will have a major impact on the learning process as a whole and not merely record its final values. This is why the prospect of a USE became a matter of hot public debate.

The same article of the Law states that assessment materials used by the USE "determine to what extent a graduate meets the federal educational standard." The existing standard of general mathematics education lays out a complex set of objectives, among them formulations such as "to foster personal culture through the study of mathematics." This principle stands alongside "mastery of mathematical skills and concepts" and other educational values.

Compliance with many (if not most) of the dictates of the standard cannot be measured directly or expressed numerically so as to form the basis of certification. The proposed USE format not only *did not* solve the problem of objective assessment; in principle, it *could not* solve it.

We can point to three specific areas where the chosen format of the USE had overtly harmed the school system.

- First of all, it caused an impoverishment of course content. The educational environment is "self-tuning" — under strict monitoring all elements not directly subject to monitoring will weaken and drop out of sight; instruction will be reduced to narrow training aimed at passing the USE, often called "drill."
- The manner in which the examination is administered is, in our opinion, immoral in a variety of ways. It fosters a perception not only of immoral individuals conducting the examination, but of an immoral educational policy, which is far more dangerous.
- Finally, the uniform examination diminishes the stature of the teacher. This has an immediate negative effect, but in time may also lead to grave consequences for the school system and the society at large — consequences that will be very difficult to correct.

7 New Challenges and New Approaches

The foregoing overview of the major events in mathematics education of the past 20 years, beginning with the dawn of *perestroika*, paints a rather gloomy picture. Having been directly involved in the events, I had always tried to take a constructive position, rather than engaging in a crossfire of arguments on the subject of the modernization of education. It has been difficult to remain perfectly neutral and to abstain from a certain critical view of the overall progress.

At the same time, the period in question was unusually fruitful in terms of the appearance of new ideas in mathematics pedagogy. Many of these ideas already have seeped into practice, though they could hardly be called widely accepted. Nevertheless, I believe this particular period will be remembered not for its specific organizational shifts (which are bound to be short-lived), but for its productive approaches.

7.1 *The Role of Mathematics in Students' Personal Development*

As noted earlier the period in question began with announcements of a shift in educational policy toward personal development. What could the study of mathematics contribute to this new direction?

This question had been debated for some time (e.g., at the XIX International Conference on Public Education in Geneva, 1956). The new principles of orientation of mathematics education were set out in the dictates of the state Standard of Education.

> The study of mathematics at the school level shall pursue the following objectives:
>
> — to impart to the student a conception of mathematics as a universal language of science and a means of modeling phenomena and processes, plus an understanding of the ideas and methods of mathematics;
> — to develop students' logical reasoning, spatial visualization, algorithmic culture, critical thinking at a level required for future professional activities and further education.

— to develop students' mastery of mathematical knowledge and skills required for day-to-day activities, the study of natural sciences and other subjects that do not require specialized training in mathematics.
— to assist in the cultivation of personality through the study of mathematics, by developing an understanding of the significance of mathematics to scientific and technological progress, and as a component of human culture through the study of the history of mathematics and the evolution of mathematical concepts (Mathematics. Course Content, 2007, p. 32).

It should be noted that, alongside the three primary objectives — psychological (cultivation of personality), utilitarian (acquisition of knowledge and skills), and pedagogic — there is a fourth "philosophical" component, aimed at the grasp of the ideas and methods of mathematics.

Clearly articulated objectives of a course of study help schools in designing course curricula; however, precisely what classwork is called for in a specific classroom, filled with specific students studying a specific topic, must be decided directly by the instructor. The function of a teacher faced with a pre-fabricated instructional system (i.e., curriculum, programs, technology, etc.) lies, on the one hand, in developing and implementing the methodology necessary to realize each instructional module and, on the other hand, in breaking down course objectives to individual lesson plans while taking into account factors such as the overall condition of the class (student group), the particular qualities of individual students, the teacher's own pedagogical preferences, etc. Several levels of planning and their corresponding objectives are included in Table 1.

Any formulation of the basic aims of general education must be accompanied by a description of expected performance. To a certain extent, existing standards of education define performance expectations for each level of study (though these may be criticized in a variety of ways). Still, despite the apparent need for clearly articulated performance expectations, actual attainment of performance levels cannot be viewed as the sole or even the primary aim of the educational

Table 1. Levels of planning.

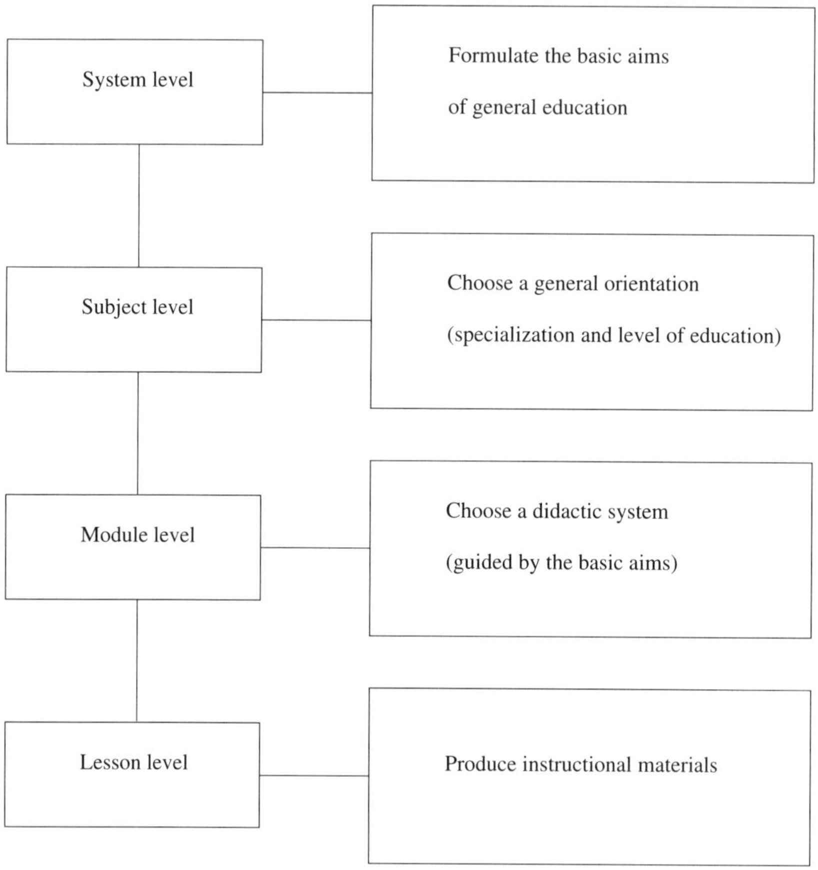

process. The Russian psychologist S. L. Bratchenko (1996, p. 97) points out that:

> the humanistic approach to pedagogy must not take for its primary aim the transmission of information or the inculcation of knowledge or skill; it is an INDIVIDUAL-CENTRIC process, the meaning and purpose of which is to FACILITATE PERSONAL GROWTH, self-knowledge, development of individual talents, broadening of personal experience, acquisition of values and improvement of understanding.

From this it can be concluded that performance requirements as defined in the state Standard must be supplemented by a variety of performance parameters, most of which cannot be expressed quantitatively. A system of such parameters was developed and published some time ago by the present author (Bashmakov, 1996). The schematic outline is given below.

Table 2. Performance parameters for mathematics.

PERSONAL DEVELOPMENT	**RANGE OF SCIENTIFIC KNOWLEDGE**		**PRODUCTIVE ACTIVITY**	
Algorithmic activity	Breadth and quality of knowledge	Calculations and transformations	**Cultivation of creative potential**	**Practical applications of reasoning abilities**
Logic and deductive reasoning	Dynamic of individual learning	Functions and graphs	Independence	Computational strategies
Visual and figurative reasoning	Depth of knowledge in selected topic	Equations and inequalities	Mental agility	Modeling
Mathematical communication and symbolic systems		Figures and solids	Capacity for generalization	Results analysis
		Measurement and geometric transformations	Adaptability to new circumstances	Self-monitoring
		Data analysis. Statistics and probability		

In this system the parameters used to evaluate performance are divided into three groups.

The first group belongs to the category of general development and describes the developmental function of education. The second group comprises more traditional criteria for evaluating performance and may be said to describe the instructional function of education. This set corresponds to the fundamental subject divisions in the study of mathematics. The third group describes student productive activity and is more complicated in structure, being intricately connected to the other two groups. The isolation of this category is especially important because it emphasizes certain aspects of student development that have suffered from lack of attention. This group may be further divided into two subgroups: the first involves the utilitarian aims of education, while the second is concerned with the cultivation of creative potential.

The principles outlined here are intended not only for abstract speculation; they may be useful in a variety of ways, e.g., in evaluating new instructional materials.

7.2 *Learning Styles*

It seems evident that any assessment of the quality and effectiveness of education ought to be based not on final test scores (the parameters in question are generally not subject to "linear ordering"), but on analyses of the educational process. Following the principles of humanistic psychology, the pedagogue is not working toward a set goal. His or her efforts are directed toward developing the students' individual talents and abilities, rather than imparting specific skills. The basic principle is that if conditions required for the PROCESS of unhampered personal development have been met, the individual will invariably attain positive RESULTS, though it may take time and effort.

For the purposes of analyzing educational processes and intellectual development, the prominent Russian psychologist M. A. Kholodnaya has been working with the category of *learning styles*. While *aptitude* describes the level of intellectual accomplishment (i.e., it is a result-based assessment), *style* is defined as a manner of performing intellectual activity (i.e., it is a process-based assessment). Consequently, different styles may result in equally high success rates in performing specific tasks.

The classification of learning styles proposed by M. A. Kholodnaya (2000) includes:

- styles of coding information;
- styles of processing information;
- styles of task formulation and problem solving.

A classification of learning styles specific to the study of mathematics has been developed in parallel with the above system; it includes the following styles:

- algorithmic
- visual
- applied
- deductive
- analytic
- combinatorial
- play.

A brief discussion of the specific nature of each style follows. The *algorithmic style* is among today's students the most common approach to clearly articulated, conventional problems. According to analyses, 80% of problems in current textbooks belong to the algorithmic (or, more broadly, reproductive) style. At the same time, we would not want to construe this style as inferior or reductive. Suffice it to say that it also includes problems that require the student to learn new, unfamiliar algorithms, select an algorithm, or modify or adapt a familiar course of action.

The term *visual style* is the most provisional of the lot. It refers to the translation of information from one language to another and the mastering of various languages, particularly the visual language. According to a widely accepted hypothesis, the study of mathematics involves three primary information languages: verbal (information presented verbally), symbolic (sequences of specialized signs or symbols), and visual (visual images and patterns). The acquisition of all three languages is the indisputable objective of the study of mathematics. Success in this field depends largely on one's ability to select the information language appropriate to a specific situation and, when necessary, to translate information from one language to another. In this respect, we should note the growing prominence of the visual language, which has led us to term this the *visual style*.

According to results collected by psychologists over the last few decades, visual language "renders meaning visible." It may be used to create visual images and manipulate them at the same level as the more conventional verbal and symbolic materials.

The *applied style* has a long and rich tradition in the Russian school system. It includes the selection of computational strategies and the solving of word problems, and more broadly, the construction and analysis of mathematical models. At the same time, practical applications of this style run into serious difficulties. The discussion of a situation and construction of a model require a great deal of time, which can seem unjustified in relation to the result. Moreover, the teacher is not always conversant with the details of the situation under discussion and he/she must spend considerable time in preparation for the class. All of these factors have reduced the use of this style to a small set of regularly presented standard situations. At the same time there is a growing need for new situations in school mathematics, particularly those involving the application of mathematics to economics, humanities, and new topics in physics.

The *deductive style* is considered to be the leading approach in the study of mathematics and traditionally is associated with geometry. The past few decades have seen significant changes in the application of the deductive style (or logic). First of all, the seemingly inviolable axiomatic structure of geometry with its clear-cut system of definitions and propositions has been deeply undermined by the educational movements that reject the necessity or even the possibility of a deductive presentation in the study of geometry; these ideas have been reflected in several new textbooks. Second, the geometrical proof has lost much of its importance in consequence of the reduced time requirements on the one hand, and the emphasis on computational problems on the other.

To a certain extent the dwindling interest in the deductive approach is linked to the separation of the course of study into algebra and geometry and the obstinate refusal to consider the possibility of their merger into a single subject. Another problem is that the study of algebra, which offers ample opportunities for the application of the deductive style, is not developed sufficiently in this regard.

The *analytic style* has lately attracted the attention of mathematics teachers thanks to the proliferation of problems "with parameters" and the widespread use of the "coordinate method." As a result, "situation problems," i.e., problems requiring extensive work within the framework of a single mathematical situation, have come to replace sequences of small unrelated problems. But analysis can be more than a "tasty side dish" to an otherwise traditional "main course." The analytic approach could take the leading role at every stage of math education and be made accessible to the average student.

If the five styles described above all are commonly used in the study of mathematics (they are distinguished, first of all, by the extent to which a style is used and the interrelations between the styles), then the following two — the combinatorial and play styles — may seem rather foreign in this context.

The *combinatorial style* implies a broad use of discrete methods and concepts: positive integers, inductive processes, and constructions, finite sets of numerical data, elements of logic, and finally combinatorics itself and elements of the theory of probability. It may seem that we are speaking of several specific topics within a mathematics curriculum. It is largely true, but with the growing predominance of digital technologies in our daily lives and across industries, we see the need to cultivate the discrete or *combinatorial* style at all levels and in all parts of mathematics education.

The possibilities of *play style* have been studied by its proponents for some time. Everyone is in agreement that games may serve as the primary means of generating interest, which leads to better understanding. But there is a distinct lack of theoretical and instructional materials that could be used to elevate play style to its full potential within the educational process.

Thus, educators must pay equal attention to all of the learning styles listed above, particularly when writing new textbooks.

7.3 *Specializations and Levels*

The most important aspect of the modernization of the Russian school system has been the shift to specialized education. This is a crucial and inevitable moment in the break with the ideology of uniform educational standards. Regrettably, official documents permit a terminological vagueness, confusing the terms *specialization* and *level of education* — two notions long distinguished in pedagogical practice. As a result, absurd formulations such as "basic level" and "specialized level" are encountered that go far to confuse educators.

The characterization of education in terms of levels is fairly widespread in both professional and lay discourse. We often hear of a *high level* of mathematics education in Russia, of a *minimal level* of education, of *advanced-level* mathematics courses, and, regrettably, of *low levels* of mathematics skills of graduates, etc. Without going into a detailed analysis of the concept of *level of education*, note that it often has an evaluative, quantitative basis. The idea is not so much in expressing this level as a concrete number (though this too is possible: if one class spends 3 hours/week on mathematics and another spends 8 hours, this yields a numerical distinction of levels), but in the possibility of comparison or linear ordering when evaluating levels.

The choice of level (often equated with the number of required clock hours) dictates the selection of materials covered by the course, the number of problems attempted, and the distribution of the types of class work. Recent practice has elicited a perception of three levels of mathematics studies: main (or standard), and two others diverging from it in either direction — the minimum level and the advanced level.

The term *specialized education* regrettably turned out to be vague. In professional training this term has a fairly clear meaning and refers to a specific profession or trade. One of the possible paths for specialized education in mathematics in fact does originate in the realm of professional training. Though specialized curricula such as "math for turners" and "math for bakers" have long been set aside, there is still a conception of specialized mathematics for major professional groups (mathematics for future economists, mathematics for communications professionals, mathematics for the humanities, etc.). Beyond the choice of content, which belongs to the category of educational level, this kind of specialization generally is distinguished by the choice of examples and problems and the inclusion (or exclusion) of specific topics. Overall, it seems that this approach is no longer relevant for secondary education.

The approach to specialized mathematics education presently emerging within the Russian school system is based on the following principle: specialization involves a choice of primary work methods, their interrelation, and their distribution. This understanding of specialization goes hand in hand with the choice of a future profession. Among the various classifications of the professional spectrum there is one that places all emphasis on the leading characteristics of professional activity, which aids in selecting the primary learning styles in the relevant study of mathematics.

The formulation of specialization, as given above, also is reconciled easily with the popular opinion that distinguishes "tracks," such as humanities, engineering, mathematics, etc. Traditional Russian courses and textbooks in mathematics may be seen as belonging to the engineering track. Until recently the Russian school was called a "polytechnic" and was expected to graduate future laborers, technicians, and engineers. This specialization is characterized by a predominance of algorithmic and constructive work methods. Visual methods are used only by way of illustration; logic and reasoning are used with extreme caution; emphasis is placed on application and subject interconnections.

The term "specialization in humanities" is not clearly defined. To most people it suggests a course of study for those who are either uninterested in or incapable of doing mathematics. A more serious understanding looks to a radical restructuring of the course of study of mathematics to emphasize its cultural component, which would require a significant expansion of visual/figurative and associative learning methods (at the expense of the algorithmic component), rethinking the nature of applications, and possibly enhancing the logic and deduction component.

The definition of specialization in mathematics education and the structuring of corresponding courses of study remain the two most significant problems of mathematics education in Russia.

7.4 *Productivity of Mathematics Education*

As mentioned earlier, freed from the constraints of the "uniform polytechnic education," Russian schools set out on the path of differentiated education. Differentiation implies, first of all, the creation of new curricula (or selection from available curricula) sometimes grounded in accumulated experience of a particular school, but more often in practical or opportunistic interests. A typical example of such differentiation is the aforementioned institution of specialized education (i.e., the creation of "tracks" in mathematics, humanities, economics, pedagogy, etc.). While many have a generally positive opinion of specialization, one emerging difficulty has been noted. Imagine that at some point along a specialized track, a student discovers that his or her course of studies has lost its appeal, become problematic or even oppressive. There was no error made in the choice of a track at the outset; but in the process of personal development (which may be especially rapid at this age) the student experienced a shift in values. Typically, that student is unable to change tracks (often this is expressly prohibited), and he or she must push forward regardless, which could lead to grave consequences. Thus, specialized education becomes for this student a veritable Procrustean bed.

Russia is in need of a flexible system of education, adaptable to the changes experienced by its students. In order for the adaptive capacities of such a system to be sufficiently rich, it must be placed within a rich environment, both in terms of funds and ideas. Not surprisingly, successful attempts to set up such systems always have extended beyond the school framework, incorporating the modern city with its resources or major national programs.

We have referred to this system as a "system of productive education," borrowing the term from Max Wertheimer and his work *Productive Thinking* (1945). Likewise, M. A. Kholodnaya (1997) speaks of the productivity of intellect. Another important work in this respect is A. N. Leontiev's *Action. Consciousness. Personality*, which develops the highly significant principle of movement (p. 185) among others and which, in turn, leads to the notion of a personal learning path:

> Human consciousness, as well as activity, are not cumulative. It is neither a plane nor a volume filled with figures and processes. Nor

> is it the connections between its discrete "units," but the inner movement of its elements, as part of the larger movement of activity that comprises the actual existence of an individual in society.

Bashmakov, Pozdnyakov, and Resnik (1997) offer the following definition of productive education:

> Productive education is a pedagogical system centered around the individual, based on a network of learning paths, which, in turn, are sequences of instructional and occupational modules independently selected by the individual student. The system makes wide use of information technology in a variety of educational applications to promote general education and culture, professional orientation and realization of the various stages of professional training, and to ensure a confident entry into society of an individual aware of his/her talents and the specifics of his/her development.
>
> The methodology of productive education presupposes:
>
> - increased student participation in the planning, realization and assessment of his/her learning path in cooperation with other participants;
> - working ties between school and workplace, school and society, school and "real life," making Productive Education into an open and flexible system;
> - the changing role of an educator as adviser and facilitator;
> - the forging of a suitable learning environment, including access to new information technologies.

The study of mathematics within the framework of productive education cannot be considered distinct from the entire educational program as realized among a specific group of students. Naturally the study of mathematics will adhere more fully to the spirit of productive education if the entire system were to undergo a restructuring. But it is also possible to implement some of the elements of productive education even in today's educational practice.

The first such element may be called *modularity of education*. Productivity of education may be increased significantly if it is offered as a sequence of relatively small (in duration and content) learning modules. This would allow for greater flexibility in the arrangement and interrelation of the modules, which is especially significant for the early stages of mathematics education, when it is still taught as a single subject, is assigned few clock hours, and encounters great diversity in the students' levels of preparation. Such an approach may be supplemented with a textbook that allows non-linear use.

A second element may be the emphasis on the *process side* of education through the introduction of a wide variety of learning materials suitable to different learning styles. Students would not be expected to attempt all available assignments: the choice of assignment will be left to the teacher and will be based on the teacher's specific objectives formulated for a specific environment. The purpose of new textbooks would be to give the teacher a sufficiently broad selection of instructional materials. There will be no complaints that some of the problems in the text are too difficult. Such textbooks would embody one of the main principles of productive education: to increase student and teacher participation in the selection of the parameters of education, including the choice of difficulty of the material and the balance of learning styles.

The third element involves *a broadening of the learning environment and enrichment of educational resources*. Here, first of all, the use of digital and web resources in parallel with (or in place of) traditional paper-based ones must be mentioned. This practice and the accompanying changes in the structuring and methodology of instruction are important future objectives.

7.5 *The Function of the Textbook*

What has been said about new resources does not mean that old resources — such as textbooks and instructional packages, which include textbooks, problem collections, and other instructional materials — therefore, are made irrelevant. On the contrary, one of the more important challenges in education is the enrichment or transformation of the function of the textbook.

In their recent work, *Psychodidactics of the Textbook*, E. Gelfman and M. Kholodnaya (2006) have given a detailed analysis of the modern-day textbook, placing special emphasis on its *informational function*. This does not imply that a textbook is a vehicle of "prepackaged" knowledge to be memorized by the student. Rather, as it is widely accepted, a textbook must function as a wellspring of cognitive problems, which the student must discover and resolve.

Speaking of the *governing function* of textbooks, the authors expressed the opinion that a serious restructuring of the form and content of traditional textbooks is demanded. In reality, the governing function of textbooks is often exaggerated. Many teachers approach the textbook as something akin to a compendium of lesson plans, and demand of it a full set of materials required for teaching a course.

The *developing function* of a textbook is least visible in contemporary publications. At the same time enhancement of motivation and the development of personal characteristics and systems of values must be at the forefront of any textbook.

One of the primary functions of today's textbook must be its *communicative function*. It is, however, often reduced to a kind of textbook-monologue. Information communicated in a textbook often is presented in telegraphic form with no regard for the complex processes that must take place in its reading.

The *pedagogical function* of textbooks generally is equated with engaging the student through discussions written in colorful language and comparisons that use striking imagery and evoke rich associations. Given the specific nature of mathematics, however, this function is better performed with lucid arguments and objective presentation.

Though today's textbooks must meet the new requirements of *differentiated* and *individualized* education, very little has been done in this respect. Individualization cannot be reduced simply to a distinction of levels of complexity; the very structure of the textbook must facilitate development of individual learning trajectories.

All of these requirements are better applied to the instructional package as a whole rather than to the textbook alone, which could hardly satisfy them all. Ignoring other types of instructional material may lead to textbooks that are overburdened and difficult to navigate.

One of the strategies for reconciling the extensive demands and the limited capacities of textbooks is to shift some of the burden onto other components of the instructional package.

7.6 *Computer-Based Technologies*

The most important challenge facing the school system over the past quarter of a century has been the prospect of introducing computer-based technologies into the educational process. This concept has been imbued with a variety of meanings because of rapid developments in the field of information technology. The basic principle underlying the varying "takes" on "computerization" has been that of an *information-based learning environment.*

> An information-based environment is a system of various kinds of interactions with human knowledge that includes storage, structuring and presentation of information (i.e. the body of accumulated knowledge), as well as its transmission, processing and further enrichment (Bashmakov, Pozdnyakov, and Reznik, 1997).

In 2000, the Russian Federation launched a federal program aimed at developing a "uniform information-based learning environment." This program has been instrumental in the building of an information infrastructure. As a result, the number of educational facilities connected to the global telecommunications network has grown dramatically. The project's aim was to effect the necessary conditions for the "computerization" of the school system, to aid in the systemic implementation and support the active use of information and communications technologies (ICT), and to orient the school system toward becoming a part of the global information community.

The following objectives had been set for the first phase of the program that concluded in 2008:

- to create a well-developed system of the production of high-quality, accessible, affordable electronic learning materials meeting the needs of students, educators, and education administrators preparing to enter into the modern knowledge economy;

- to broaden and enhance professional training options aimed at the implementation of ICT in educational practice;
- to create district-wide methodological centers in participating districts, offering support in "computerization" and implementation of new teaching practices.

The current phase of the program addresses the following goals:

- to spread the use of digital educational resources developed during the first phase of the program throughout most schools;
- to introduce nationwide ICT-based professional training courses for educators, aimed at higher pedagogical performance and helping educators explore information technology as an instrument of their continuing professional development;
- to share accumulated experience in facilitating computerization and using ICT in methodological support;
- to develop production of high-quality educational materials and enter the international market of digital learning resources.

During the first phase, mathematics instruction was enhanced with specially created "new-generation" teaching materials. This resulted in an effective system of producing, implementing, disseminating, and maintaining digital teaching resources. An important role in this process is played by professional development courses that train educators to use ICT in their work and in improving the educational process.

Work is underway to create a system of district-wide methodological centers, which could provide support in the "computerization" of schools, disseminate newly developed teaching materials, and offer professional training for educators and school administrators. These centers will be connected virtually, thus forming a uniform computer-based learning environment.

In the course of the program there has been a rise in the production of innovative educational packages that make use of computer technologies in various aspects of the educational process, from algorithmic tests to research support. At the same time ongoing development of complexly structured information resources that offer access to deep information environments, from handbooks and encyclopedias to complex didactic games, is underway.

If during the first phase of the project the lack of equipment and technology was deemed to be the main obstacle to its advancement, today the gap in pedagogical planning and insights needed for the creation of environments capable of fostering students' creative potential are considered more important.

7.7 *The Role of Academic Institutions in Renewing Educational Content*

If the system of formal education of the past 150 years is analyzed, it will be noted that significant changes in content occur every 30–40 years at best. These shifts are dictated by various factors typically rooted in social changes. It is far more difficult to trace the influence of scientific advancement on the subject matter of public education, though such influence over large periods of time is indisputable. There has been a great deal of discussion on this subject between prominent scientists and educators. Consider this excerpt from an article by the famed French mathematician Émile Borel (2000, p. 30)

> Astounding scientific breakthroughs of the 18th c, which had brought about the technological developments of the 19th c., are tied to four great figures: Galileo, Descartes, Newton and Leibniz. There is neither a single physical object nor a single idea that does not bear the imprint of the scientific revolution of the 18th c. There is no single sphere of human activity that has remained unaffected by the genius of one of the four men. But here I have misspoken: something has indeed evaded this influence and remains largely unchanged; I am referring to the system of secondary education.... Someone may object that it appears dangerous to tie formal education to scientific progress.... Should we not be concerned that new sophisticated subjects that have not been properly adapted by schools would be less beneficial for general education than the tried and true ones?

The first topic of discussion concerning the influence of academic institutions on formal education should be that of translation of the major achievements of modern science into school curricula. Occasionally, this idea makes it through the bureaucratic sieve, and then topics such as laws of genetics, set theory, foundations of relativity theory, programming languages, etc., are hastily introduced into school programs. Universities become involved in the writing of textbooks and the training or retraining of teachers. After a short time (approximately 6–10 years) there is a backlash, because it turns out that teachers are having difficulties incorporating new topics into traditional curricula, that students emerge from the classroom with a set of new terms and no grasp of the substance, and that the new textbooks are difficult and poorly written. Public opinion is rallied in defense of children against the evils of innovation (it is sufficient to recall the blackballing of mathematics textbooks at the Congress of the Supreme Soviet in the 1970s); finally everything goes back to the old ways, though a distinct positive trace of the innovations remains, especially among progressive educators.

During the same period other processes are taking place with far graver consequences. In the final analysis, they amount to no more than a bowdlerization of educational content by getting rid of fundamental scientific ideas and replacing them with simplified formulas and "rules of conduct." This process is tied to a far greater extent to government policy and is more difficult to analyze over any significant period of time and across national borders.

Consider, for example, mathematics education in Russian schools today. It virtually has been purged of a great number of classical notions (e.g., symmetry, induction, number theory, geometric transformations, a few types of constructions, etc.). Much of class time is spent in mastering a narrow set of solving skills in accordance with prescribed formulas (e.g., finding domains of equations, solving standard type inequalities or exponential and logarithmic equations, etc.). The practical use of these skills in future professional activity (to say nothing of the purported claims of personality development) is so vulnerable to criticism (especially expert criticism) that ensuing calls for reduced mathematics requirements are readily heeded and quickly implemented. The sole remaining argument for maintaining the present teaching style is that it prepares students for college entrance examinations. Needless to say, this system is flawed and unsustainable. The Russian school system is falling prey to forces of illiteracy and ignorance.

Fortunately, universities are far less susceptible to this trend. As a result, the primary role of universities in structuring formal education must be concerned with the creation of a uniform educational system, where the universities (not government agencies and methodological entities) will dictate and maintain educational content. The excesses that may occur with this arrangement (such as overburdened curricula, overestimated learning abilities, etc.) are far less harmful than the official policy of stunting the intellectual growth of the young generation by isolating them from the achievements of world culture.

Universities have their own ways of influencing the education system (these deserve a separate discussion), and it is their aim to use this influence in the service of long-term goals and prospects, rather than that of government policies.

The principal enduring influence of universities on the school system occurs through the production of instructional and popularizing texts. As far as textbooks are concerned, the current Russian climate is largely favorable, though the path a new text must take through the wilderness of various commissions and expert councils remains troublesome. It should be noted, however, that the current circumstances are not being utilized by universities to their fullest potential. Apparently there are not enough incentives in place to attract leading scientists to the production of textbooks.

It should be pointed out that the influence of universities on the school system in the 1960s–1980s had a distinct social component alongside the purely pedagogical one. At that time, academic institutions were refining methods of instruction that aimed at instilling in the average student an interest in the sciences. Regrettably, universities are losing the broad social basis that had served as a solid foundation in training future scientists, and the inevitable consequences of this development will be felt in the very near future.

8 Compared with Other Countries

The problems and challenges discussed above are not unique to Russia; therefore, it may be useful to compare the state of affairs in Russia with that abroad. Over the course of the last few decades there have been several large-scale international comparative studies of mathematics education. One such study, in which I had the honor of participating, will be discussed below.

In 1998, the European Union asked the European Mathematical Society (EMS) to sponsor a study entitled "Comparative features of levels of education for students 16 years of age." In this work members of the EMS commission for mathematics education, which also included the present author, were joined by 10–12 international experts from specific countries. Although the study was oriented toward the countries of the EU, there were several experts on the commission from countries that did not belong to the union, including Russia. It should be noted that the commission had very high regard for the state of mathematics education in Russia and had made extensive use of Russian materials and documents, which were assembled (alongside materials from other countries) at the specially created information center in Besançon (France).

The findings of the study were discussed at a subsequent conference, which gathered not only mathematicians, but also high-level public officials from all of the EU countries. The materials resulting from the commission's work included the general report, a list of mathematics problems used in the comparative study, and individual national reports. They can be found online at http://pegase.univ-fcomte.fr/ctu/IREM/Internat.htm.

It should be noted that the objectives of the study, as well as the presentation of its findings, are radically different from those of the seemingly analogous international projects that had taken place over the last few decades (such as TIMSS, PISA, etc.). Unfortunately, as noted by all the participants, the title of this project did not match its aims, since they had little to do with evaluating or comparing the levels of education across various countries or, even less, with potential recommendations for its standardization. The key principle of the study had been:

> The achievements of mathematic education across the European countries, including its national traditions and distinctions, are a universal treasure, which must be understood, preserved and developed.

We should note also that the age of 16 specified by the study was chosen as the average age at which mandatory education is terminated in the majority of the countries.

8.1 *How Are We the Same?*

What is it that unites mathematicians of all nations in their ideas on mathematics education?

1. The aims of mathematics education are understood to be more or less identical, despite the differences in their formulation, which may be traced in the assembled official documents (curricula, national standards, etc.). The most important principles uniting all the participants may be articulated in the following manner:

The definition of a general course of studies in mathematics has always comprised two competing tendencies: the utilitarian (pragmatic), geared toward the practical applications of mathematics in daily activities, and the conceptual, concerned with promoting the role of mathematics in the general development of human beings. If the social conditions of the second half of the 20th century had favored the utilitarian approach, the recent changes in these conditions command a definite shift toward the conceptual tendency, which will continue to grow in the immediate future.

2. Ideas concerning the role of mathematics in the general development of human beings have turned out to be very similar. It was pleasing to see the outline of parameters characterizing this role, as it had been developed in Russia, accepted as the European standard practically without significant changes.

3. Despite all the differences in evolved pedagogic traditions, experts had no trouble finding common ground. This was true of a wide

range of issues, including course content, professional training, aptitude assessment, the place of mathematics in general education, etc. In the past the methods of teaching mathematics have had a pronounced national flavor tied to national curricula and textbooks. More recently there has been a marked internationalization of the field. Despite formal obstacles, countries are actively sharing educational materials.

4. All of the countries are experiencing increasing pressure to carry out school reform in general and mathematics reform in particular. The need for reform is linked to a number of larger societal changes (e.g., the rise of information technologies, growing social stratification, general tendencies toward standardization, etc.). At the same time all members of the Commission agreed that no substantial changes to the content and methods of instruction can be made before its objectives and perspectives are clarified and widely accepted.

The Commission emphasized the need to resist actively all ill-formulated attempts at "standardizing" education, which cause more harm by ignoring many important aspects of human development than they do good in their supposed pursuit of equality and accessibility.

8.2 *How Are We Different?*

An analysis of individual national reports has shown that educational systems adopted across Europe are significantly different from one another. Some of the most striking differences include:

1. The terms of mandatory education vary widely from country to country, from 14 to 18 years, which makes it difficult to evaluate performance at a specific age (e.g., at 16, as the Commission chose to do for the project in question). The Commission noted, however, a general tendency toward an increased term of mandatory training.

2. The point at which education becomes specialized differs as well. Several countries (Austria, Belgium, Holland, UK) begin specialized studies as early as 11–12, or even 10 in some parts of Germany, whereas in Russia, Finland, and Sweden the split is not made until 16–17. Nevertheless, a general tendency to postpone specialization of education was noted, which often goes against established practice.

3. In the course of the Commission's investigations significant differences in the choice of primary methods of instruction were discovered. Speaking very broadly, a number of countries adhere to the traditional approach of "transmission of knowledge," while others have resolutely adapted the "constructive" method, where the process of knowledge acquisition is considered more important than the totality of knowledge itself.

4. The number of required clock hours varies widely across the curricula of various countries: e.g., requirements for students of 16 years of age range from 3 to 8 classes per week. Despite the general tendency toward reduced mathematics requirements, the subject still maintains a second place on the curricula of most countries (behind native language and literature).

5. There are great discrepancies in the usage of computers in teaching mathematics. This is a complex issue, subject to rapid changes, and for this reason we shall not deal with it at this time.

6. The most interesting finding of the Commission's study proved to be the difference in subject matter of mathematics and its distribution across grade levels. Five topics in mathematics were selected for comparison:

Quadratic equations
The Pythagorean theorem
Similarity
Percent
Word problems.

The varying approaches to the teaching of these topics across Europe may be gleaned from the national reports posted on the Commission's website. One interesting point is that in several countries (France being the most prominent example) the study of substantive topics is postponed as much as possible. As a result, the final two years of mathematics are seriously overburdened. Consider the following example: the administrators of the international competition, Kangaroo, run into the same difficulty in the selection of problems year after year because topics such as the Pythagorean theorem, polynomial factorization, inequalities, etc., are taught later in Europe than they are in Russia.

Naturally, there are many other discrepancies in the teaching of mathematics across various countries, such as the extent of innovative processes, the structure and nature of textbook use, the role and structure of examinations, professional training for teachers, etc.

An important component of the international investigation was a set of problems that allowed the Commission to consider the distinct parameters of mathematics education and the manifestation of important mathematical ideas in course curricula. There was a total of 65 problems, which were posed to representatives of various countries. Each problem was accompanied by a brief "bio," stating the problem's origin, purpose, the mathematical idea it embodies, level of difficulty, and recommendation for use.

It should be noted right away that the Commission has stressed, both in the accompanying text and in various discussions, that these problems were meant to provide "food for thought" rather than offer a platform for comparison. One of the most active members, Rudolf Sträßer, has written that "the worst possible outcome of the international comparative test would be a ranking of the European countries, the division into champions and second-tier countries."

The Commission recommended that the problems (which are available online) be used to analyze national traditions and emerging trends. They may serve as indicators of weak and strong aspects of mathematics education in a given region or a given group of schools.

9 In Place of a Conclusion

My thoughts on the history of mathematics in Russia and the problems facing it, as well as my informal discussions with European mathematicians about questions of mathematics education, have allowed me to formulate some ideas with which I would like to conclude my overview.

1. The world is made more beautiful through the pursuit of mathematics, and therefore it is a joy, an honor, and a responsibility to be in a society of people who occupy themselves with transmitting this adornment to future generations. The positive emotions, the enjoyable aspects of our work make up for the difficulties we have to face in our lives.

2. It is crucial to maintain a balance between the two tendencies: to preserve the traditional core of mathematics education on the one hand, and to renew the content and the methods of this education on the other.

As an ardent proponent of the first of these tendencies, I would like to point out that the so-called core is not sharply delineated. This circumstance opens the floodgates of speculative attacks on any and all changes and novelties. We must examine our attitude toward these changes not in relation to "what we were taught" or "what we had taught" with perfectly good results, but in relation to the overall set of objectives and subject matter of the study of mathematics.

3. We must be more resolute in ridding ourselves of formalism in our teaching practices. First priority must be given to the grasping of essences of mathematical ideas. Even when we speak of fostering logical thought or accuracy of expression as important objectives in the study of mathematics, we should keep in mind that these qualities must be developed through the mastery of complex ideas and materials.

4. The greatest challenge facing mathematics education is that of evaluating student performance. To this day our copious guidebooks are filled with all manner of problems designed to measure levels of education. Moreover, naturally-arising demands for increased or decreased volumes of mathematics instruction for this or that group of students lead to frequent modifications of this list. It is naïve to think that a student will be able to solve many of these problems (or even understand their formulation) just a few years after graduation. All of these circumstances lead to a very limited and impoverished set of problems presented in the classroom, especially where required clock hours for mathematics have been cut. We believe that another model deserves attention, one where students (no matter what the level or credit requirement) are presented with a substantive, engaging, and full-scale course of study, without expectation of future reproducibility. Experience gained in this type of classroom may prove more important (and be retained longer) than the drilling aimed at the performance of basic tasks. Naturally this approach requires serious methodological preparation, but we have already seen examples of its practical applications.

References

Bashmakov, M. I. (1991). *Algebra i nachala analiza* [*Algebra and Elementary Calculus*]. Moscow: Prosveschenie.

Bashmakov, M. I. (1996). Planirovanie uchitelem svoey deyatelnosti [Planning your teaching activities]. Vestnik SZO RAO, Obrazovanie i kultura Severo-Zapada Rossii, vyp. 1, St. Petersburg, 140–145.

Bashmakov, M., Pozdnyakov, S., and Reznik, N. (1997). *Informatsionnaya sreda obucheniya* [*Information-Based Learning Environment*] (pp. 4–5). St. Petersburg: Svet.

Borel, E. (2000). Kak soglasovat' prepodavanie v sredney shkole s progressom nauki. [Reconciling secondary education with scientific progress]. In *Matematika v obrazovanii i vospitanii.* Moscow: Fazis.

Bratchenko, S. L. (1996). Gumanisticheskoe napravlenie obrazovaniya [Humanistic education]. Vestnik SZO RAO, Obrazovanie i kultura Severo-Zapada Rossii, vyp. 1, St. Petersburg.

Dneprov, E. D. (1994). *Chetvertaya shkolnaya reforma v Rossii* [*The Fourth School Reform in Russia*]. Moscow: Interpraks.

Dieudonné, J. (1972). *Lineynaya algebra i elementarnaya geometriya* [*Linear Algebra and Elementary Geometry*]. Moscow: Nauka.

Formulation of Federal Standards of General Education. (2008). [*Kontseptsiya federalnykh gosudarstvennykh obrazovatelnykh standartov obschego obrazovaniya*]. Moscow: Prosveschenie.

Fridman, L. M. (1987). *Pedagogicheskii opyt glazami psikhologa* [*Pedagogy from a Psychologist's Perspective*]. Moscow: Prosveschenie.

Gelfman, E., and Kholodnaya, M. (2006). *Psikhodidaktika shkolnogo uchebnika* [*Psychodidactics of Textbooks*]. St. Petersburg: Piter.

Kholodnaya, M. (1997). *Psikhologiya intellekta: paradoksy issledovaniya* [*Psychology of Intellect: Paradoxes of Investigation*]. Moscow: Bars.

Kholodnaya, M. (2000). *Formirovanie personalnogo poznavatelnogo stilya uchenika kak odno iz napravlenii individualizatsii obucheniya* [*Development of a Personal Learning Style as a Way Toward Individualized Education*]. Shkolnye technologii, #4, Moscow: Narodnoe obrazovanie.

Krylov, A. N. (1956). *Vospominaniya i ocherki* [*Recollections and Essays*]. Moscow: AN SSSR.

Legislative Acts of the Russian Federation Pertaining to Education [*Zakonodatelnye akty Rossiiskoi Federatsii v sfere obrazovaniya*]. (2008). Moscow: RAO.

Leontiev, A. N. (1983). *Izbrannye psikhologicheskie proizvedeniya* [*Selected Works in Psychology*], Vol. 2. Moscow: Pedagogika.

Mathematics. Course Content [*Matematika. Soderzhanie obrazovaniya*]. (2007). Moscow: Ventana-Graf.

Monakhov, V. M. (1983). *Problemy edinogo urovnya obscheobrazovatelnoy podgotovki uchaschikhsya v srednikh uchebnykh zavedeniyakh* [*Problems of Uniform Standards in Secondary Schools*]. Moscow: Pedagogika.

Nurk, E. R., and Telgmaa, A. E. (1988). *Matematika 4–5. Uchebnik dlya sredney shkoly* [*Mathematics. A Textbook for the 4–5 grades*]. M.: Prosveschenie.

A Recommendation of the Conference to Ministries of Public Education Concerning Mathematics Education in Secondary Schools [Rekomendatsiya konferentsii ministerstvam narodnogo prosvescheniya, otnosyashayasya k prepodavaniiu matematiki v srednikh shkolakh] (2000). In *Matematika v obrazovanii i vospitanii* (pp. 53–55). Moscow: Fazis.

Vavilov, N. A. (2003). *Ne sovsem naivnaya teoriya otobrazhenii* [*A Not Entirely Naïve Theory of Mappings*]. St. Petersburg: MMF SPBGU.

Wertheimer, M. (1987). *Produktivnoe myshlenie* [*Productive Thinking*]. Moscow: Progress.

5

Mathematicians and Mathematics Education: A Tradition of Involvement

Alexey Sossinsky

Independent University of Moscow, Russia

1 Introduction

One of the most remarkable characteristics of Russian school-level mathematics education is the fact that research mathematicians have participated in it in a variety of different ways. World-famous mathematicians such as A. N. Kolmogorov and A. D. Alexandrov have written school textbooks. Hundreds of research mathematicians (including the two just mentioned) have spent many hours teaching or observing classes in schools. Many Russian research mathematicians have become passionately involved in organizing mathematics clubs (or "mathematics circles," as they are known in Russia) or in writing problems for all kinds of mathematics olympiads.

The aim of this chapter is to describe several examples of such service to mathematics education. Such a description, naturally, will be limited — the history of the involvement of Russian mathematicians in the problems of school-level education might begin with Euler, and certainly with Chebyshev, but I will focus mainly on what happened after 1955. I do this for two reasons: first, 1955–1956 was a watershed of sorts in Russian politics — soon after Stalin's death, a mild liberalization (with crucial psychological and social effects) was implemented by Nikita Khrushchev; and second, because it was in

1955 that the author of this chapter first became acquainted with the mathematics teaching activities in the Soviet Union. Also, I have restricted the geography to present-day Russia (and so to the Russian Federation of Soviet times) with some reluctance: there were many interesting people and activities in Kiev, Minsk, Tbilissi, and Batumi that I would have been happy to mention, but this would only be a small part of a general picture that I am not qualified to describe. Nor can I claim to be able to give an objective overall picture of what happened in provincial Russia. I was personally involved in many of the activities described below, but mostly in Moscow and (to a lesser extent) St. Petersburg (Leningrad), and my information about what went on in other Russian cities is sporadic and incomplete.

But even within these chronological and geographic limitations, there are many objective difficulties in treating the topics of the present chapter, especially concerning Soviet times. Since the most interesting activities were in a sense marginal, taking place, as it were, outside the control of the Soviet pedagogical establishment, since mainstream publications about them were few and far between, and since many of the people involved are no longer with us, at present there are very few sources of objective information. Thus many of the descriptions below are based on the author's personal recollections and those of several colleagues whom I was able to contact in the preparation of the present text. The author has tried very hard to be objective, still, the text below does not pretend to be anything more than an active participant's personal sketch of a fascinating aspect of mathematics education in Russia in the second half of the 20th century. Concerning my own involvement in the activities described below, see Sossinsky (2007).

For the sake of convenience, in describing mathematics teaching in Russia (Soviet Union) one can divide the period in question into three parts: a relatively liberal period, which roughly corresponds to the Khrushchev years, although it came to an end only in the late 1960s, after Khrushchev's resignation; the period of stagnation or the Brezhnev era; and the period that began with "perestroika," in 1985 and continued in subsequent years. The Khrushchev period was characterized by the implementation of numerous initiatives related to mathematical education, in particular by the development of the olympiad movement, the resurgence of mathematics lectures for the general public and high-school students, the proliferation of mathematics circles, the creation of specialized schools, the appearance of the popular science magazine for high-school students *Kvant* (Quantum), the success of I. M. Gelfand's Mathematics Correspondence School, and the appearance of summer institutes. A measure of the remarkable development of interest in mathematics and theoretical physics, boosted by the launching of the first sputnik, was the immediate success of *Kvant* magazine, which was created in 1969 by the nuclear physicist I. K. Kikoin and the great mathematician A. N. Kolmogorov. Among other noteworthy and influential events, the 6th and 10th International Mathematics Olympiads that took place in Moscow and Leningrad in 1964 and 1968 should be mentioned.

The stagnation period was a period of return to totalitarian principles, which led to serious problems for the semi-official structures created in the 1960s. Not everything went smoothly, as exemplified by the crackdown by authorities on some of the specialized schools (e.g., School #2 in Moscow), the partial purge of the Kolmogorov boarding school (School #18) of "politically unreliable elements," and the total replacement of the Jury of the National Olympiad (mainly consisting of the people who had created it) when it was taken over by the Ministry of Education in 1979. Nevertheless, in the Brezhnev years, although few significant new developments occurred, the wide network of extracurricular activities, specialized schools, and competitions continued to function rather efficiently despite the political and economic difficulties.

Finally, the advent of perestroika, which was a much more fundamental liberalization than the "Khrushchev thaw," is characterized by the appearance of new structures (e.g., MCCME, see Section 11), new types of competitions, and the continuing functioning of the structures created in the 1960s with varying degrees of success.

2 Mathematics Circles and Olympiads

Let us begin by focusing on the traditional ways in which mathematicians have worked in mathematics education: by organizing mathematics circles and olympiads. Mathematics circles and olympiads appeared in Russia as early as the 1930s, and it is not my purpose here to trace their entire history. But no discussion of the role of mathematicians in the education of future mathematicians during the period in question can fail to mention their influence.

A "kruzhok" (literally, a little circle) is an unofficial gathering of people who get together, usually once a week, for discussions about a topic of common interest, be it science, literature or politics. Such circles were usually headed by a charismatic expert, whose aim was the dissemination of knowledge, or culture, or political propaganda (as in the famous Marxist circles that eventually created a small but strong social power base for the bolsheviks among industrial workers). As a rule, in tsarist times, the circles functioned outside the official establishment and most of the political ones were illegal.

Of course, mathematics circles for high-school students in the 1950s and 1960s had no connection with politics (with some notable exceptions, see below), but it is important to understand that, from the organizational point of view, they were based on enthusiasm and had little connection with the existing Soviet educational establishment. None of the people running the circles were paid for doing it, the only support from universities and high schools being the classrooms usually put at their disposal during weekends. It should be stressed that the vast majority of mathematics circles were conducted by university professors, instructors, graduate students, and even undergraduates. It was unusual for ordinary secondary school teachers to conduct a mathematics circle; the exceptions were very few and far between. One should also mention, as a modified version of a mathematics circle, the general mathematics seminars officially intended for university freshmen, in which some advanced high-school students also participated. In confirmation of this fact, I might point out that at the beginning of A. Kirillov's (1972) famous and quite advanced book *Elements of Representation Theory*, Kirillov notes the necessity of conveying fundamental necessary knowledge to high-school students who might be involved in studying the topics discussed in the book.

Boltyansky and Yaglom (1965) have painted a vivid picture of the life of Moscow's mathematics circles, describing traditions in whose creation they themselves played a considerable part. The history and practices of St. Petersburg's (Leningrad's) mathematics circles was described by Fomin, Genkin, and Itenberg (1996). Arguably, these circles were a leading factor in the creation of one of the most remarkable generation of Russian research mathematicians, that of Manin, Sinai, Arnold, Dobrushin, Fuchs, Novikov, Kirillov, Vinberg, and others. Many outstanding mathematicians in their turn later returned to mathematics circles, now as leaders rather than students.

In the 1950s and 1960s, there were so many first-rate mathematics circles in Moscow alone, that a sufficiently complete list (or even an incomplete but representative one) would fill pages and pages. Certainly, it is a fact that the overwhelming majority of future scientists, whether doing fundamental research or engaged in engineering, computer science and information technologies, data processing, cryptography, design of the nuclear bombs, space rockets, military hardware, participated in at least one math circle in their teens. Numerous successful mathematics circles functioned in the 1970s and later, but their role in forming the succeeding generations of mathematicians was probably not as important because, by then, new pathways to mathematics, e.g., specialized city schools and math–physics boarding schools, had appeared.

It is widely known that most (in fact almost all) the best Russian mathematicians born after 1930 participated in olympiads (and usually earned first prizes), so such a list would be long and would practically coincide with that of all the best Russian mathematicians of those generations. It is easier to list some exceptions: Yu. Manin, M. Ratner, R. Minlos, A. Okounkov. Although the Fields medalist (the Fields medal is the highest award in mathematics, a kind of Nobel prize) S. P. Novikov claims that he was never interested in olympiads, I have not included him in this brief list because actually he did earn a second prize at the prestigious Moscow olympiad when he was in eighth grade (but never participated again).

In Russia, the Olympiads were first organized in St. Petersburg (then called Leningrad). Foremost among their organizers was the geometrician Boris Delaunay (Delone), a charismatic, extravagant, and versatile personality, a great story-teller, known not only for his work

Boris Delaunay

in research mathematics but for his exploits as a mountain climber. In 1935, he moved from Leningrad to Moscow and from that time on took an active part in organizing olympiads in the capital as well, serving for a time as the chairman of the Moscow Olympiads Organizational Committee. This position was occupied by other major mathematicians as well, among them A. N. Kolmogorov and V. I. Arnold.

The 6th and 10th International Mathematics Olympiads, which took place in Moscow and Leningrad in 1964 and 1968, played an important role in the development of the olympiad movement in Russia. As the years went by, the importance of olympiads became more and more apparent not only to educators, but also to bureaucrats of the Ministry of Education and even to politicians. In 1979, the olympiad movement, which had been totally independent was "swallowed up" by the ministry (more about this below).

3 M. A. Lavrentiev, A. N. Kolmogorov, D. K. Faddeev, and the Math–Physics Boarding Schools

In the mid-1950s, the organization of secondary schools in Soviet Russia, and, in particular, the teaching of mathematics, was quite stable.

Schools were strictly controlled, highly centralized, with uniform nationwide programs and prescribed textbooks, the level of math teachers was relatively solid and the level of secondary school students was quite high.

At that time, a number of Russian scientists, beginning with Nobel Prize winner N. N. Semenov, realized that the further technological development of the Soviet Union required well-trained specialists in the mathematical sciences, and that this training should begin at the secondary school level, the most talented students being brought together in elite schools (see Abramov, 2009). Of course, the word "elite" was never used in this context, being in contradiction with the egalitarian communist ideology. But elite schools, especially musical schools and schools with an emphasis on studying a foreign language, the latter usually catering to the children of high-ranking communist party members (those who were "more equal than others," to quote Orwell's imperishable phrase), were already in existence, and the hypocritical communist phraseology used to justify their creation was successively used in the creation of specialized math schools.

The first of these specialized math–physics boarding schools was founded in Akademgorodok, the Siberian scientific center that arose near Novosibirsk in the early 1960s. Initially, in 1962, it operated effectively as a correspondence school; classes first began at the school in January 1963. The school was organized by the mathematician M. A. Lavrentiev, the "Tsar of Akademgorodok" with the help A. A. Lyapunov, one of the fathers of Russian cybernetics. A special math syllabus for the school was created, and the teachers were all university people, including young instructors and graduate students. The academicians Lavrentiev and Sobolev, and Soviet Academy of Sciences corresponding members Lyapunov and M. M. Lavrentiev, all taught courses at the school, which frequently lasted for as long as two years.

The math–physics boarding School #18 in Moscow was founded by Isaak Kikoin (the well-known nuclear physicist, one of the fathers of the Russian A- and H-bombs) and the mathematician Andrey Kolmogorov in 1963 (Abramov, 2008). The students were preselected each year via the olympiads; the best performers were brought together in a month-long summer school where the final selection was made. Mathematics at School #18 was taught by Kolmogorov himself and several young mathematicians, including outstanding ones like V. I. Arnold and V. M. Alekseev. The syllabus was developed under Kolmogorov's guidance, as was the original list of mathematics teachers, only two of whom, A. A. Shershevsky and I. K. Surin, were "ordinary" school teachers (not university research mathematicians). Their role was to make the pupils of the school proficient in the kind of elementary mathematics that was needed for the entrance exams to MSU. Kolmogorov's role in the creation and development of School #18 was enormous. His vast work in ordinary schools is not addressed in this chapter (see Chap. 3), but his participation in the life of School #18 was highly important as well. The personality of A. N. Kolmogorov is discussed in many sources, in particular Tikhomirov (2007) and Arnold (2007).

Most of the school's alumni successfully passed the entrance examinations to Mekhmat, the Mechanics and Mathematics Department of MSU, and a number of them eventually became outstanding research mathematicians. Among the members of the first three graduating classes of School #18, A. Abrashkin, A. Arkhipov, I. Krichever, L. Levin, Yu. Matiyasevich, S. Matveev, V. Nikulin, S. Voronin, and A. Zvonkin should be mentioned; all of them are internationally recognized leaders in their respective fields of research (for example, Yu. Matiyasevich solved Hilbert's tenth problem, I. Krichever is the chair of the mathematics department at Columbia University, S. Matveev is a corresponding member of the Russian Academy of Sciences, etc.). It is noteworthy that some of these individuals returned to their alma mater for at least brief periods of time as teachers, and that three students who graduated from the school during its first three years — A. Abramov, A. Zemlyakov, and V. Dubrovsky — became not only teachers there, but important contributors to Russian mathematics education in general.

The same year, a similar school, the Leningrad Boarding School #45, was founded in Leningrad as the result of the efforts of several professional mathematicians (including D. K. Faddeev, A. D. Alexandrov, and A. A. Nikitin) and several of their much younger Leningrad University colleagues (in particular, M. Bashmakov, Yu. Ionin, and A. Plotkin).

The outstanding algebraist, D. K. Faddeev, officially headed the newly created School #45 as the chairman of its Board of Trustees.

The famous geometer A. D. Alexandrov, then rector (i.e., president) of Leningrad University, a versatile personality who was, like Delaunay, a top-level mountain climber, gave the school whatever help he could. A. A. Nikitin, a mathematician with a military background, very influential in high-level Leningrad party circles, was very helpful politically. Other Leningrad mathematicians were later also of great importance for the school. The great researcher in topology, algebraic geometry, and ergodic theory V. A. Rokhlin (about his fascinating life story, see Vershik, 1990) frequently discussed the mathematics syllabus with the school faculty. Another research mathematician who did a great deal in syllabus reform for specialized schools was the geometrician V. A. Zalgaller.

It may be argued, however, that the central figure in the creation of School #45 and its first 10 years of operation was the then-young M. I. Bashmakov. He was born in 1937 in Leningrad. Bashmakov became interested in mathematics in high school and was an assiduous problem-solver in the city's leading mathematics circle (headed by G. Epifanov). Mark was always an extremely energetic and outgoing person, a great traveler and, again, a mountain climber. He successfully

Mark Bashmakov

used his energy as a Komsomol (Young Communist League) leader, not on the ideological front, but by using the Young Communist League's structures to organize different useful activities related, in particular, to mathematics education. At the same time, he was a talented research mathematician (PhD in 1963, Doctor of Sciences in 1974 in algebra, in particular K-theory) and an efficient olympiad problem composer and solver. Already in his freshman year at the university, he headed a mathematics circle of his own, became active in the organization of the Leningrad olympiad and, eventually, in the All-Union olympiad. Bashmakov was responsible for all practical aspects of the school's creation.

It is interesting to mention that subsequently Bashmakov became involved in an area of mathematics education that is seemingly very distant from mathematically gifted education. In 1970s, Romanov, the ambitious head of the Leningrad party organization, decreed that 40% of the high-school students of Leningrad should spend the last two or three years of their secondary education in vocational schools. Bashmakov began teaching mathematics in one of these vocational schools, was involved in the mathematics syllabus reforms in these schools and wrote several textbooks for them (see, e.g., Bashmakov, 1987).

4 Israel Gelfand, the Correspondence School and Seminar

One of the greatest and most prolific mathematicians of the 20th century, Israel Gelfand, also was deeply involved in mathematics education. Born in 1913 near Odessa in a modest Jewish family, his education was far from ordinary: he was accepted in the mathematics graduate school at Moscow State University at the age of 19 without any university diploma and, in fact, even without a high-school diploma, and became a pupil of Kolmogorov, immediately obtaining brilliant mathematical results. Today Israel Gelfand is professor emeritus at Rutgers University, where, despite his age, he continues his mathematical research.[1]

[1] Editor's note: Israel Gelfand died in October 2009 while this book was in preparation.

As a 21-year-old docent (associate professor) at Moscow State University, Gelfand already supervised mathematics circles, but his main educational achievement was the creation, in 1964, of the famous VZMSh, a nationwide mathematics correspondence school, whose aim was to help provincial teachers and high-school students (from schools far away from university centers) learn mathematics and prepare for university entrance examinations. Gelfand, with the help of a dedicated team of university graduates and some other mathematicians (including A. A. Kirillov, a brilliant research mathematician, now at the University of Pennsylvania), worked out the principles by which the school functioned, specified its syllabus, and wrote textbooks for it (see, for example, Gelfand, Glagoleva, and Kirillov, 1973). The school was very successful and efficient and it established strong ties between Gelfand's highly qualified team and the best provincial teachers, in particular via what was then called a "collective pupil."

Gelfand's main organizational achievement, of course, was not VZMSh, but his own mathematical seminar at MSU, probably the biggest and best mathematical seminar of all time. It was a research seminar for university mathematicians, but it must be mentioned here,

Israel Gelfand

because it has a non-trivial relationship with the subject matter of this chapter.

From the end of the 1950s and until Gelfand's departure to the US in 1990, the seminar gathered in room 1408, an auditorium with a seating capacity of 150, which was always almost full, and lasted 4 hours or more. It would begin as soon as Gelfand arrived, which could be at any time between 7 and 8 pm (however, participants who came late, i.e., after its leader, were not allowed in). Basically the seminar was a one-man show: Gelfand would sit or stand facing the participants and often interrupt the speaker, asking him questions or commenting on the talk (other participants were not allowed to do that).

The best mathematicians and theoretical physicists of Russia (and occasionally from abroad) participated and gave talks at the seminar about the most important recent developments in their respective fields, but often Gelfand's commentary was more interesting and enlightening than the speaker's talk itself. Gelfand, however, could destroy a talk by an eminent expert at the very beginning by interrupting it, telling the audience that the speaker's approach was basically flawed, very convincingly explaining why, concluding that the seminar could not waste its time listening to the rest of the talk, and then calling on the next speaker on the agenda to begin the report to follow.

Does all this have anything to do with high-school math education? It does! During most of the time of the seminar's functioning, the first two rows of the left-hand side of auditorium 1408 were filled with very gifted high-school students, some of whom Gelfand knew by name, and would address, interrupting the speaker, to explain some important point that he suspected the students did not understand.

Let me recall a more dramatic instance of the participation of secondary school pupils in the seminar, that occurred in the early 1970s. After the first 20 minutes of a talk by a well-known middle-aged mathematician (we will call him "Professor Ivanov"), often interrupted by Gelfand's aggressive questions, Gelfand suddenly declared: "Let's stop right here." He went on to say that "Ivanov's" methods for studying the subject of the talk were absolutely inappropriate, and, addressing one of the high-school students in the first row (we will call him "Misha"), explained why and what the correct approach should be, frequently asking "'Misha', do you follow me?", and concluded by saying that "Misha" would give a talk on the same subject the next week and Professor "Ivanov" was invited to the seminar to learn what the correct approach to the subject should be. Indeed, the very next week, apparently coached by Gelfand, "Misha" gave such a talk, and "Ivanov" was there to listen. Perhaps it should be added that today "Misha" is a world-famous mathematician, and that this story illustrates one of Israel Gelfand's remarkable traits: the ability to recognize talents and direct them, albeit by unusual means, to specific research topics.

After his arrival in the US, Gelfand did not forget his involvement in Russian high-school mathematics education. In a letter written to the alumni of School #2 (see about the school below) on the occasion of the school's 50th anniversary (and published in Koval'dzhi, 2006, p. 218), Gelfand wrote the following:

> I would like to note four important traits which are common to mathematics, music, and other arts and sciences: the first is beauty, the second is simplicity, the third is precision, and the fourth is ... crazy ideas.

Frankly, I know no better definition of high class mathematics.

5 A. S. Kronrod, E. B. Dynkin, and the City Selective Mathematics Schools

Approximately at the same time as the math–physics boarding schools were being organized, a number of mathematicians decided to help create selective mathematically-oriented schools in the major cities.[2] In some cases the main motivation of these mathematicians were their children: they wanted them to study at high-level schools emphasizing the mathematical sciences.

The first of these schools in Moscow, School #7, was organized by Alexander Kronrod, a charismatic person and brilliant mathematician, whose independence and freedom-loving attitude never particularly

[2]Editor's note: Mathematics classes began to appear even earlier — in 1959, under the supervision of S. I. Shvartsburd.

pleased the authorities. In 1962, several of the classes of what used to be an ordinary city school, #7, were designated as "mathematical classes." These classes had a few extra hours of mathematics each week, and mathematics lessons were conducted by university people, including Kronrod himself (one of the pupils was Misha Kronrod, Alexander's younger son). N. N. Konstantinov (more about him below) also taught there. The reputation of the school surged immediately, and entrance examinations had to be organized each year to select its future students among the numerous applicants. Upon graduation, practically all the alumni of these math classes took the entrance exams to Mekhmat MSU, with a remarkably high percentage of success. But the director of the school was not too happy with the students and teachers of the mathematics classes, who were not easily managed. After Kronrod and some of his colleagues lost interest in the school and were not replaced by teachers of the same caliber, the level of the school declined and it was no longer capable of competing with other selective mathematical schools that appeared in the 1960s.

One of the most famous was School #2, located in the South-West outskirts of Moscow, not far from the new MSU campus. It began functioning in 1956, while the construction of its standard five-story building was still being completed. The first director, Vladimir Ovchinnikov, was a history teacher, a former Komsomol (Young Communist League) leader who progressively became critical of the communist regime, combining the courage needed to stand for his convictions with the intelligence required to avoid direct conflict with the authorities. One of the first key steps in organizing the school was his choice of "professional training" of its pupils. One of Khrushchev's educational reforms required that each school should train its pupils in some trade, preferably in one of the local enterprises. Instead of the local shoe factory, Ovchinnikov reached an agreement with the director of one of the Academy of Sciences research institutes located nearby, and the school's pupils were trained in various research-related tasks at that institute. Another key step was hiring several first-rate teachers, mostly in the humanities.

But the main step in the transformation of School #2 into an elite math school was taken when Israel Gelfand and Evgeny Dynkin (an outstanding research mathematician, now professor emeritus at Cornell) became involved in the teaching of mathematics there in 1963 and 1964, respectively. This was the result of clever maneuvering on the part of Ovchinnikov, who agreed to accept their children to his school provided that their famous parents would teach there and/or help organize mathematical classes. Among the teachers of mathematics classes at School #2 were Alexey Tolpygo (a young Ukranian mathematician and great olympiad problem-solver, then doing his PhD at MSU), Boris Geidman (a brilliant high-school teacher, presently the author of primary school math textbooks), and, later, Sergei Smirnov (then an MSU student doing a PhD in topology, now better known for his "problem books" in Russian history) and the future dissident Valery Senderov, then a young graduate of the prestigious Physico-Technical Institute.

Gelfand and Dynkin both lectured at the school, were active in syllabus development, and helped recruit young university mathematicians to do most of the math teaching. In a sense, the level of teaching and the atmosphere at School #2 by the end of the 1960s was too good to last (for more details, see Koval'dzhi, 2006, and indeed a crackdown was soon to occur (see below).

Outside Moscow, selective math–physics schools or schools with math classes sprang up in most of the bigger cities. In Leningrad, Schools #30, #38, #239 should be mentioned. (Grigory Perelman, who was awarded but refused the Fields medal in 2006, graduated from the latter in early 1980s.) I reluctantly omit mentioning those math–physics schools I know in other cities, although many of them were first-rate.

6 Nikolay Vasiliev and the *Kvant* Magazine Problem-Solving Section

One more side of Kolmogorov's involvement in secondary school mathematics education, in addition to what was mentioned above, was the popularization of mathematics via the journal *Kvant*. Here again Kolmogorov cooperated with the physicist I. K. Kikoin. The journal, which targeted bright high-school students interested in mathematics and physics, was launched in January 1970, with an original circulation of 100 000. The circulation rose to the incredible level of 350,000 per month in 1972, and remained stable around the level of 200,000 up to the first years of perestroika.

Besides Kolmogorov himself, the authors included famous mathematicians like S. Fomin, Fuchs, Gelfand, Kirillov, M. Krein, Pontryagin, Rokhlin, Tikhomirov, Viro, and Zalgaller. Younger authors (mainly former olympiad winners) as well as rank-and-file school teachers also contributed articles. The mathematical part of the journal was edited by the algebraist L. Makar-Limanov and the author of this chapter. The task of editing sometimes amounted to rewriting articles from beginning to end. One of the most successful sections of the journal was the problem section for younger readers, headed by the topologist Anatoly Savin (with marvelous illustrations by E. Nazarov). The regular publication of problems (with solutions) proposed at the entrance examinations of the leading Russian universities played a considerable role in boosting the readership. The entire collection of *Kvant* from 1970 to the present day is available at the website <www.mccme.ru/kvant>.

But here I would like to dwell on another section of *Kvant*, the monthly math problem section headed by Nikolai Vasiliev. Kolya Vasiliev, a soft-spoken and very musical person, had hoped for a career as a concert pianist. It is only as a high-school senior in 1960 that he realized that music would not be his profession. As a student of a musical school, Vasiliev was good in mathematics but did not participate regularly in any circles or in Olympiads. To his own surprise, he succeeded in passing the difficult entrance exams to Mekhmat MSU. There it progressively became clear that he was an incomparable solver of difficult olympiad problems. He soon became one of the central figures of the Moscow Olympiad and, later, of the All-Union (national) Olympiad, and later still, of Konstantinov's Tournament of Towns as a composer of beautiful challenging problems. It was this taste for elementary mathematical challenges that Vasiliev brought to the magazine *Kvant*.

Every month, from January 1970 to his untimely death in 1998, Vasiliev would select four problems for the mathematics problem

Nikolai Vasiliev

section of that month's issue of *Kvant*, and oversee the publication of their solutions in the next issue. The readers still in school were encouraged to send their solutions to the magazine and thereby to participate in the *Kvant Problem Solving Contest*. The solutions were corrected and, at the end of the academic year, the best problem solvers received prizes. Many of the prizewinners eventually became scientists. I will only mention the names of four well-known research mathematicians: R. Bezrukavnikov (now professor at MIT), I. Itenberg (professor at the University of Strasbourg), I. Arzhantsev (associate professor at MSU), and G. Perelman (recently granted the Fields medal, which he did not accept).

The math problems of the *Kvant Problem Solving Contest* are collected in two booklets (see Vasiliev, 2005–2006). Some of them appeared in English in the issues of *Quantum* magazine, the American version of *Kvant*.

7 A Few Words on Humanities

This chapter and this whole book are devoted to mathematics and mathematicians. But it is impossible to describe the life of the mathematics community and its involvement in education without touching on the humanities as well — without this, the overall atmosphere would remain unclear. Several examples should suffice. Ovchinnikov, the director of School #2, has already been mentioned. Among the school's teachers was Anatoly Yakobson, a gifted writer and literary critic and one of the editors of the *Chronicle of Current Events*, an underground "samizdat" journal. In the words of Radio Liberty, "Yakobson was one of the true heroes of that time" (broadcast of 27 September 2008). Yakobson taught literature and history, and skipped the obligatory communist propaganda in his lessons. In his classes, and those of other literature teachers, readings of writers not appearing in the official syllabus, e.g., Bulgakov or even Solzhenitsyn, or out-of-favor poets like Akhmatova or Mandelshtam, were not unusual. Ovchinnikov was certainly aware of this and, as the school's director, should have put an end to such practices, but pretended not to notice them. In the mid-1960s, the pupils and teachers were very appreciative of the liberal atmosphere of the school, and were careful to avoid having their lessons branded as "anti-Soviet" and thus provide the authorities with a pretext to crackdown on the school. In 1968, when KGB's interest in Yakobson's activities became too serious, Yakobson quit his position at School #2, explaining to the director that it would not be in the school's interest to have one of its teachers arrested as an anti-Soviet dissident. Later he was forced to immigrate to Israel, where in 1978 he committed suicide. During that same year, A. D. Sakharov nominated him along with seven other Soviet dissidents for the Nobel Peace Prize (http://www.antho.net/library/yacobson/about/andrei-sakharov.html).

The director of Kolmogorov's School #18, Raissa Ostraya, was in many respects similar to Ovchinnikov. She was a graduate of the prestigious but short-lived IFLI (Institute of Philosophy, Literature, and History), a woman who had proved her courage as a front-line nurse during the war (during which she had joined the communist party), but found it hard to reconcile her high cultural level and respect for the liberal attitude of the teachers, especially the young teachers of mathematics, with the demands of the party's ideology and educational policy.

One of the literature and social sciences teachers was Yuly Kim, the famous singer-songwriter and playwright. Kim had ties to dissidents' circles and signed a number of letters that expressed oppositional views. From 1968 on, he no longer worked at School No. 18. As the *Chronicle of Current Events* reported, he had been "fired 'on his own volition', all of his musical performances were cancelled, and his contract to play the leading part in a film was torn up."

The *Chronicle*, however, also listed the names of many mathematicians, including mathematicians connected with education, who had suffered for their excessively liberal views, which clashed with the new spirit of the times.

8 Tightening of the Screws. How Did Mathematics Education Fare?

In 1964, in a well-organized and bloodless coup, Leonid Brezhnev succeeded in ousting Nikita Khrushchev and becoming the head of the communist party and of the USSR. In 1968 the Soviet Army invaded Czechoslovakia, putting an end to the Prague Spring. This was a fundamental change of policy, a return to a more totalitarian regime. As was to be expected, attempts were made to tighten the screws on the liberal acquisitions in mathematics education that appeared in the 1960s, such as Kolmogorov's syllabus reforms, the selective schools, the national olympiad movement, the math–physics boarding schools, the magazine *Kvant*. Fortunately, none of these attempts succeeded in completely destroying these structures although some, especially in Moscow and Leningrad, went through difficult periods.

Perhaps the first important event took place at MSU, where the liberal administration and party organization were replaced by hard-liners after the appearance of the so-called "Letter of the 99" — an open letter of protest against the forcible incarceration of the well-known dissident and logician A. Esenin-Volpin in an insane asylum by the KGB in 1969 (see Fuchs, 2007). Soon after that, the anti-semitic practices at the Mekhmat entrance exams (which had ceased after Stalin's death) were resumed.

On the secondary school level, 1970 was a year during which most of the directors of the Moscow high schools with special mathematics classes were fired, and in some cases the special mathematics classes were discontinued. The most dramatic instance was perhaps that of School #2, where the local party authorities accused the administration and the teachers of all kinds of ideological wrongdoings. A special commission was created, and its investigations resulted in all the best teachers, as well as the director, Ovchinnikov, being fired and mathematics classes suspended.

Things were not much better in Leningrad: Schools #38 and #30 were merged into one (in other words, one of the schools was shut down), which was moved far away from the center of the city. School #121 was eliminated.

In Kolmogorov's boarding School #18, after Yu. Kim had circulated a brash political open letter, the MSU party organization, without organizing an overt investigation (as in School #2), decided to clean out the "undesirables" progressively, which included the director, R. Ostraya, and most of the mathematics teachers originally selected by Kolmogorov (the author of this chapter was one of the first to go). The school never returned to the high educational and cultural level that had characterized it in the 1960s, but remains a first-rate training center for future students of the mathematical sciences.

As part of the anti-Kolmogorov campaign (see Chap. 3), Kolmogorov's opponents also headed a takeover bid of *Kvant* magazine, but it did not succeed due to the strong and intelligent stand taken by Kikoin (see Sossinsky, 2007, p. 240).

The Mekhmat party leaders were very unhappy about the success of Gelfand's correspondence school, but could not find a pretext to attack it. Instead, they organized another such school, called "Malyi Mekhmat," to compete with VZMSh, and they planned to progressively entice its pupils to switch from the Gelfand school to theirs. To this end, they were promising their pupils a preferential treatment at the entrance exams. But this enterprise failed: Gelfand's VZMSh was so much better than Malyi Mekhmat, that after a few years the latter was discontinued for lack of students, while VZMSh is still going strong in Russia today.

Another important event along the same lines was the change of status of the National Olympiad. In 1979, it lost its independence and was taken over by the Ministry of Education. That year the 13th All-Union Olympiad took place in Tbilisi, and for the first time its organization was supervised by a bureaucrat from the Ministry. That person immediately ordered that important changes be made in the entire procedure, in complete contradiction with existing traditions. The informal and friendly atmosphere that had characterized the previous olympiads did not suit her. She regarded the contestants and the Jury as potential cheaters, and devised all sorts of rules to eliminate all contacts between them, refusing to believe that no one had ever attempted anything dishonest at the National Olympiads, and that for many of the provincial participants contacts with the professional mathematicians from the Jury were their first opportunity to talk with a real research mathematician. However, the Jury succeeded in disregarding these rules: on the last day of the Olympiad, the correction of the papers and contacts with the contestants proceeded as they had in previous years, while the lady from the Ministry was touring Tbilisi wine cellars, accompanied by one of the local organizers, a young and handsome Georgian. Quite inebriated, she had not dared appear until the closing ceremony, so that the 13th Olympiad was conducted according to tradition.

But this was a Pyrrhic victory for the Jury: all but two of its members (including this author) were dismissed by the Ministry, never to participate in the National Olympiad Jury again. The Jury of the 14th and subsequent olympiads was constituted by highly qualified, but more manageable people. The excellent mathematical level remained, but the friendly and informal atmosphere, as far as I know, never returned.

9 Bella Subbotovskaya and the Tragedy of the "People's University"

During the period of stagnation, the very participation of a mathematician in mathematics education could be seen as a political deed, for which people sometimes paid with their lives. One such tragic episode from the annals of Russian mathematics education will be described below (also see Szpiro, 2007).

A short-lived unofficial institution was known under various names: the ironic title "People's University" or the more descriptive (but nevertheless misleading) "Jewish Seminar," but I prefer calling it the "Bella Muchnik School" by the name of its founder, Bella Subbotovskaya-Muchnik. In spirit, this school was closer to a mathematics circle than to a bona fide university, and an important part of it dealt with high-school students. Bella, an alumnus of Mekhmat MSU, worked as a high-school teacher after obtaining her PhD (a very unusual career choice for someone with those credentials). One of her tasks in school was preparing talented students for the entrance exams of her alma mater, which after 1968 had become aggressively anti-semitic and, more generally, biased against applicants from selective Moscow schools. One of her helpers was Valery Senderov, one of the mathematics class leaders at School #2 (mentioned above). In the late 1970s, Bella decided that bright students flunked at the exams (for being Jewish, or part Jewish, or too smart) should get a chance to get as good a mathematical education as the successful applicants. (About anti-semitism at the Mekhmat entrance examinations, see the study "Intellectual Genocide" by V. Senderov and B. Kanevsky in the book Shifman, 2007.)

To do this, she "recruited" her MSU classmate, the topologist Alexander Vinogradov, who began teaching there, together with some of his pupils and participants of his seminar (including the present author) on a biweekly basis. The first year (1980–1981) "classes" took place in Bella's cramped apartment, and in the second semester the "staff" was joined by Dmitry Fuchs, Andrey Zelevinsky, Alexander Shen, Arkady Vaintrop, and some other mathematicians (at present mostly known for their outstanding research work), willing to take the risk of teaching at an institution of doubtful ideological correctness. The mathematicians mentioned above were not dissidents and taught only mathematics, carefully avoiding "anti-Soviet" discussions, although Valery Senderov, besides teaching mathematics at Bella's school (mostly to high-school students preparing for entrance exams), often gathered some of the students after classes for discussions about politics and religion.

The "enrollment" was typically 40–50 people (about half as many by the end of the first year), most of whom were undergraduates (mostly not from MSU!), but some were still high-school students. In 1981, classes were transferred to a classroom in the Gubkin Oil and Gas Institute, acquiring the semi-official status of optional seminar for undergraduates of that institution. The students were also learning mathematics at the different places where they were officially enrolled, so that the curriculum in Bella's school was not traditional — basic modern mathematics was taught there in a novel way. Apparently, the courses were often more interesting than their counterparts at MSU, and some undergraduates from Mekhmat (not only freshmen) actually attended. A memorable event in the period when the classroom was still a room in Bella's apartment was a superb lecture by the American Fields medalist John Milnor, then in Moscow for a short visit.

The Bella Subbotovskaya school ended its existence under very dramatic circumstances, in the winter of 1982–1983. V. Senderov was arrested, as well as two students (one was soon released). Several

Bella Subbotovskaya

students were interviewed by the KGB. Classes ceased of their own accord. Soon afterwards, Bella Subbotovskaya, returning home very late, was run over by a truck. The police never found the hit-and-run driver. Few of the people involved believed that it was an accident. Classes never reopened.

In 2007, a conference in memory of Bella Subbotovskaya was held in Haifa.

10 Nikolai Konstantinov

Nikolai Konstantinov, who recently celebrated his 77th birthday, is best described as the living legend of Russian mathematics education. Most of his organizational and teaching activities took place in totalitarian Russia, but, as the freethinker that he was, Konstantinov managed, for several decades on the borderline of Soviet legality, to create various extremely efficient and successful semi-official structures, that were all about learning and doing mathematics. He remained active and productive during the new *perestroika* period and during the post-perestroika period as well. Therefore, his story will help to bring this chapter to the very recent past and to the present day.

Born in Moscow in 1932, Nikolai Konstantinov became interested in biology at an early age (in particular in Darwin's theories and in genetics) and under normal circumstances would have certainly become a biologist. We owe the fact that he is a great mathematics educator to the odious figure of T. D. Lysenko. The sadly famous academic showdown of August 1948 in which Lysenko succeeded in destroying the promising school of genetic biology in Russia (see Soifer, 1994) was attentively followed by Konstantinov (still in high school), and it became clear to him that the study of genetics was no longer possible in Russia. That autumn he took and passed the entrance examinations to the Physics Department (his second best choice) of Moscow State University.

As a mathematician, Konstantinov is a product of the mathematics circle system. In high school, he was a participant in a small mathematics circle headed by V. A. Uspensky (the future logician and pupil of Kolmogorov, then a freshman at Mekhmat). The other participants included Mikhail Agranovich, Felix Berezin, and Robert Minlos — all of them eventually becoming famous researchers in partial differential equations and mathematical physics. At the Physics Department, under the influence of A. S. Kronrod and A. A. Lyapunov, Konstantinov was more and more attracted to mathematics and, after graduation from the department, obtained an assistant professor position at the Mathematics Chair there (then headed by the Academician A. N. Tikhonov).

Before returning to the mathematics circle system for high-school students as a teacher, and in his fifth and last year as an undergraduate at the Physics Department, Konstantinov tried his hand at teaching mathematics (functional analysis) to Mekhmat freshmen and sophomores in the framework of a semi-official seminar. "Alumni" of that teaching experience include Nikolai Rozov, Eugene Golod, Alexander Venttsel, and some other research mathematicians.

It is only in 1960 that Konstantinov, then 28, organized his first mathematics circle for high-school students, which he called "Kruzhok Alpha." Among its regular participants were Joseph (Ossya) and David (Dodik) Bernstein (both brothers were International Mathematics Olympiad (IMO) winners in subsequent years), Volik Fishman, Alexander Geronimus and other mathematics researchers-to-be. The future Fields medalist, Grigory Margoulis, was an occasional visitor. Later there was a "Kruzhok Beta," mathematics classes in Kronrod's School #7, in Schools #57, #91, #179, summer mathematics institutes in Estonia, a leading role in the organization of the Moscow Mathematics Olympiad and permanent membership in the Jury of the All-Union Mathematics Olympiad until 1979, when almost all of the Jury were dismissed, as discussed above.

Konstantinov's reaction to this event was quite simple: "Since I have been kicked out of the official olympiad, I will create my own unofficial one." In 1980, he organized the "Tournament of the Three Cities," which became a national event under the title "Tournament of Towns" the next year and an international one (ITT) a few years later, with Bulgaria and Australia the first countries to join the bandwagon. The organizational principles are brilliant in their simplicity: it is a mathematical problem-solving competition between cities; small and large towns participate on an equal basis due to a democratic formula which takes into account the town's population. The competition is held twice a year on the same Sunday with the same set of problems, supplied to the local organizers by Konstantinov's Moscow team. The papers are corrected locally, and the best ones are sent to Moscow for review. The cities are ranked in accordance with the "democratic formula" mentioned above, the best individual participants are invited to the ITT Summer School.

The remarkable thing is that the total expenses for running the ITT are negligible, as compared, say, to those of the IMO. There are no travel expenses; all that is needed for the local organization is one dedicated university mathematician (perhaps with a few enthusiastic graduate students willing to help grade the papers) and some small mailing and advertising costs! In Moscow, the program committee, headed by Konstantinov, also functions mainly on enthusiasm, so that it manages with the revenue that it gets from the almost nominal participation fees of the cities. For a more detailed account (in English) of the ITT, including the problems (with solutions) from 1980 to 1997, see the four booklets edited by P. J. Taylor (Taylor, 1992, 1993, 1996, 1998).

An important addition to the Tournament of Towns are the ITT Summer Schools, which are completely different from the Tournament of Towns itself: there are no problem-solving competitions in limited time, no first, second, etc. prizes awarded, no olympiad-style mathematics. What the school tries to do is to give an imitation of mathematical research, to involve the students in a two-week-long model of how professional mathematicians work. The school is headed by N. N. Konstantinov and its mathematics program is run by his younger collaborators. To my knowledge, there are no publications in English about the summer schools, but those who read Russian may refer to Frenkin (2009), which contains a selection of the best series of problems discussed in the school.

One should not think that the principal goal of these schools is to recruit future research mathematicians, since not very many are needed anyway. More important, to my mind, is the fact that many successful mathematics olympiad performers learn what research in pure mathematics is really like, and begin to determine correctly what

N. N. Konstantinov

the role of mathematics should be in their future careers — perhaps not the main object of study, but simply an important tool, or just another tool (along with the computer).

The ITT and its Summer Schools are by no means Konstantinov's only organizational achievement. He is, among other things, the principal founder of the IUM (Independent University of Moscow, founded in 1991), of the MCCME (Moscow Center of Continuous Mathematical Education, founded in 1997), and of the Lomonosov Competition (1979).

The Lomonosov Competition is multidisciplinary and targets "average" secondary school students (and not only those from elite schools), including pupils just out of middle school. It was first held in 1977, gathering over a thousand pupils in several auditoriums at various Moscow institutions of higher learning, where the participants were offered a wide choice of subjects — not only problem solving in mathematics, physics, and chemistry, but competitive activities in biology, history, literature, psychology, mathematical games, linguistics, etc.

I was one of the organizers of the mathematical games competition, in which 12–14 year old pupils first played certain logical combinatorial games (new to them) with each other, discussed strategy with the instructor, and, eventually, tried to understand what a winning strategy is and whether it exists for the given game. When the day's work was over, Konstantinov asked me what I thought about the competition and whether I understood what the point of the whole thing was. I said something vague about getting youngsters interested in topics beyond the standard school material, but Konstantinov explained that what I said was of course true, but I had missed the main organizational point altogether — the main idea being that of the database, which contained absolutely invaluable information, in particular for people intending to organize extracurricular activities. Thus a psychologist intending to set up a psychology circle for teenagers could easily get the addresses of all the participants who visited the psychology activities of the Lomonosov competition and send them letters advertising the circle. As many of Konstantinov's creations, the underlying ideas are simple and extremely efficient.

10.1 *Moscow School #57*

Among the educational institutions connected with Konstantinov one must mention School #57, which remains to this day arguably one of the best schools in the world. Its "mathematization" began in 1970, when N. N. Konstantinov was asked to organize a mathematics class there by a local high-school administrator named Bogdanova, one of the rare occasions when an educational bureaucrat was the initiator of an elite high school specializing in mathematics. Actually Konstantinov worked there for only one year, and the flourishing of School #57 began later, after the mid-1980s. As in the case of School #2, the director, N. E. Lapushkina, played a very positive role.

Among the people who taught mathematics there at various times were Victor Vassiliev, Alexander Shen, Victor Ginzburg, Maxim Kontsevich (all of whom are world-class research mathematicians, the latter is a Fields medalist), Boris Geidman, Rafail Gordin, Boris Davidovich, Ivan Yashchenko (university graduates with research experience who became great high-school teachers). Not long before the advent of perestroika, the local and city party authorities decided that the liberal atmosphere in the school was "anti-Soviet" and that a crackdown on the school was in order. They replaced the principal, held responsible for "ideological errors," with S. L. Mendelevich, who was to head the "clean-up." But the result was highly unexpected for them. The new principal liked the atmosphere and the teaching process so well that he decided not only to keep the best teachers on the staff, but invited new university-level mathematicians, in particular, Boris Davidovich, who became the school's assistant principal in 1986. The school flourished, and still does. The reader who understands Russian can consult the school's web site www.sch57.msk.ru.

10.2 *The MCCME: A New Structure and New People*

The MCCME is another one of Konstantinov's creations. The main pretext of its organization was to install legally the IUM (about the IUM, see Sossinsky, 1997) in a new building given to it by the prefect of the Central District of Moscow, A. I. Muzykansky. This could not be done directly, because under the existing rules city authorities can legally support education only at the school level, not on the university level. And so the building was given to the MCCME, a newly created independent organization involved in mathematics education in primary and secondary schools, but a sizable part of the classrooms in the building were in fact allotted to the IUM.

The MCCME has, since then, become an extremely efficient center of mathematics education, running and supervising mathematics circles, olympiads, and other mathematical competitions, as well as summer institutes, assisting the computerization of high schools, organizing teacher training and teacher contests, publishing mathematical books, creating mathematics education databases, housing Konstantinov's team that runs the ITT, and the Lomonosov Tournament, and managing the financial and practical affairs of the IUM. The Center has an excellent relationship with the Moscow Department of Education, which finances many of its projects and activities. The MCCME is headed by two remarkable administrators in their early 40s, the director Ivan Yashchenko, a PhD from MSU with teaching experience in elite schools and mathematics circles, and the financial director Victor Furin, also a graduate of Mechmat (who also has a university diploma in law). They head a team of well-organized, young but very competent underpaid enthusiasts who regard the dissemination of mathematical knowledge as a noble cause and a way of life. At present, the role of the MCCME in Russian mathematics education is much more important than that of the IUM, for the benefit of which the Center was originally created.

Although one of the Center's main functions is to coordinate mathematics competitions, challenges, specialized schools, and math classes, thus catering to the elite, this does not contradict the Center's deep involvement in mathematics education of ordinary pupils in ordinary schools, and, as a consequence, in teacher training, in the mathematics syllabus and in mathematics textbooks. Confining myself to one example of the Center's activities, I will describe a project aimed at advanced participants — the Dubna Summer School on "Contemporary Mathematics."

It is held around 5 km away from the physics research center Dubna, around 120 km from Moscow, at the place where the Dubna river flows into the Volga, in a nice vacation facility surrounded by pine forests. The school is only nine years old. It takes place in the last two weeks of July, bringing together high-school juniors and seniors, university freshmen and sophomores (80–100 students in all), and a sizable group of distinguished research mathematicians and university teachers.

Most of the students are current or former olympiad prizewinners, but some places are reserved for students without such distinctions, who apply to the school via the Internet. They are accepted if the organizers appreciate how they have filled in the school's rather unusual application forms, in which they are asked to describe, in brief essay form, their interest in mathematics, answering questions such as: What was the last mathematical book that you have read and how did you like it? or: What mathematical proofs are your favorites (present two)? or further: What mathematical constructions have most impressed you? The recommendations of teachers, especially teachers of selective schools, are also taken into account. Each year since 2007, the Dubna gatherings have hosted 8–12 foreign students (from Western Europe) with a special program (lectures in English or translations of the most interesting Russian lectures).

The highlight of the school are lectures by famous mathematicians, mathematical superstars like D. Anosov, V. Arnold, P. Deligne, Yu. Ilyashenko, S. Novikov, Ya. Sinai, V. Vassiliev, E. Ghys, and Yu. Matiyasevich. There is also a wide choice of cycles of three-four exercise classes by younger mathematicians, either on the material of the lectures or on other topics. These classes (unlike the lectures) take place in parallel, so that the instructors compete for the listeners. Overall, the program is perhaps overloaded, but then the students are totally free to visit whatever courses they like and are, in fact, encouraged not to take too many. There are no examinations, no mathematical competitions of any kind (but of course the students are given problems to solve both in the lectures and in the other classes).

An important feature of the school, expressed in its title ("Contemporary Mathematics") is that the topics discussed in it are chosen so as to lead up to the frontiers of present-day research. As yet unsolved problems are often presented at the end of the lectures or courses. The aim, of course, is to demonstrate that mathematics is a living science, still intensively being created, and not a rigid body of knowledge that one must learn and then apply, as mathematics teachers often tend to believe and explain to their pupils.

Participants are asked to write a little essay about their impressions. In their assessment of the school, many of the participants stress that only now have they begun to understand what doing mathematics is, that it is not only solving problems previously solved by others in limited time and memorizing theorems and proofs — it is a creative process.

The school has developed stable traditions in non-mathematical activities: informal discussions, swimming, sports (there are always football, volleyball, table tennis tournaments), poetry readings, the traditional boat ride on the Volga, and campfire songs. An important feature of all of these is the accessibility to the students in an informal atmosphere of professional mathematicians (including the superstars). Thus the Dubna school not only shows what mathematics is all about, but also what working mathematicians are like as people. More information about the school (part of it in English) is available at the website www.mccme.ru/dubna.

11 Mathematics Teachers and Research Mathematicians. Friends or Foes?

In discussing the tradition of professional mathematicians' involvement in education, it would be wrong not to mention the fact that their relations with "ordinary" teachers were not always unproblematic. I think it can be said that many teachers in Russia were not very happy with university people (in particular young research mathematicians) working with their pupils, and did not support mathematics competitions, olympiad-type problems, and other challenges. Why?

First of all, some Russian teachers dislike mathematical challenges of all kinds, because they feel that challenges put their authority in question. Being unable to solve challenging problems (such as those from the *Kvant* problem section, which some of their students do solve) weakens the teacher's position of respect by the students. Unlike professional mathematicians, who readily admit that they do not know how to solve a given problem immediately and are very happy when one of their students finds a solution faster, some "ordinary" teachers are reluctant to admit this inability, regarding it as a professional shortcoming.

Moreover, in the period under consideration, many mathematics teachers in Soviet Russia were unfamiliar with the material (usually not included in the standard school syllabus) that university people usually taught to their pupils. If the teachers came to such classes, they were often shocked by the total disregard for pedagogical and didactical principles characterizing the teaching style of university people (especially young ones), while observing that the latter's lessons attracted the gifted students much more than their own (by the sheer enthusiasm of those giving them and/or by the intrinsic interest of the new material). Some teachers were also wary of some of the anti-establishment remarks that young mathematicians would sometimes make, regarding them as dangerous and/or demagogical. On the other hand, it must be admitted, of course, that many of the young university instructors or graduate students who worked with gifted high-school students tended to be rather arrogant in their contacts with "ordinary" teachers, even good ones, lacking an adequate appreciation of the specific requirements and difficulties of their job. To illustrate this fact, I will quote from a book by Zvonkin, a graduate of School #18 and currently a professor at the University of Bordeaux. In describing how he spent a whole month working as an ordinary first-grade teacher in one of Moscow's schools, he notes:

> Prior to this, I was a self-assured intellectual, always ready to criticize schools and teachers and to offer them wise advice. This cruel, but extremely useful experiment with myself forced me to change many of my views (Zvonkin, 2007, pp. 12–13).

Of course, the two communities (teachers and mathematicians) did not always distrust each other. For example, the relationship between the two "ordinary" math teachers at Kolmogorov's School #18, Shershevsky and Surin, and the university people teaching mathematics there, was that of mutual respect and cooperation. It should be also noted that although olympiads were originally organized by university mathematicians, as their pyramidal structure (school, district, city, republic) developed, they were ordinarily conducted at the school level by the local school teachers. In this context, these teachers often cooperated with university people, who helped, in particular, in the selection of problems for the school-level competition. This was another example of successful cooperation between mathematicians and educators.

Fortunately, I think that it is fair to say that the situation has been evolving in a positive direction since perestroika, partly because of the more relaxed political atmosphere and also because of the appearance of new types of competitions in which rank-and-file teachers are actively involved. Another reason for increased cooperation between teachers and mathematicians are the examples of university mathematicians whose research career was aborted or abandoned in favor of high-school teaching. There were many such successful transformations.

12 Conclusion

The text above was meant to be a tribute to all the dedicated people, be they academicians, university professors, research mathematicians young and old, or "ordinary" mathematics teachers, who overcame the difficulties of Soviet and post-Soviet times to create the 20th century Russian mathematical school. I must apologize to those I know (or knew) for not having included their names or mentioned them only in passing, and to those whom I never met, but who are perhaps no less important and deserving, for not having researched the subject deeply enough to write about them.

Let me stress that the half-century-long story sketched above cannot be understood outside the context of the Russian mentality and cultural traditions. The success of the structures and activities that I have described can be understood only within that context; their descriptions are not recipes for success in mathematics education that can be automatically implemented elsewhere. Nevertheless, it is fair to say that the people I talk about have not only worked for Russia, but have succeeded in producing an important contribution to the international mathematical scene. The main results of their efforts are highly-qualified researchers and teachers, now dispersed in mathematics departments at universities all over the world. And some of them have been successful in carrying over part of the Russian (or should I say East-European?) traditions to their host country.

Acknowledgments

I am grateful to Nikolai Konstantinov, Andrey Egorov, Mark Bashmakov, Alexander Abramov, Ivan Yashchenko, Valery Senderov, and Boris Davidovich for generously sharing their recollections with me.

References

Abramov, A. (Ed.). (2008). *Kikoin, Kolmogorov, FMSh MGU* [*Kikoin, Kolmogorov, and the Math–Physics High School at Moscow University*]. Moscow: Fazis.

Abramov, A. M. (Ed.). (2009). *A. N. Kolmogorov: publitsistika* [*A. N. Kolmogorov: Publicistic Works*]. Moscow: MTsNMO.

Arnold, V. I. (2007). On A. N. Kolmogorov. In S. Zdraskovska and P. L. Duren (Eds.), *The Golden Years of Russian Mathematics* (pp. 129–153). Providence, R.I.: AMS Publications.

Bashmakov, M. I. (1987). *Matematika: eksperimintalnoe uchebnoe posobie dlya SPTU* [*Mathematics: Experimental Textbook for Vocational Schools*]. Moscow: Vysshaya Shkola.

Boltyansky, V. G., and Yaglom, I. M. (1965). Shkolnyi matematicheskii kruzhok pri MGU i moskovskie matematicheskie olimpiady [The school math circle at Moscow State University and the Moscow mathematics Olympiads]. In Leman, A. A., *Sbornik zadach moskovskikh matematicheskikh olimpiad* (pp. 3–46). Moscow: Prosveschenie.

Fomin, D., Genkin, S., and Itenberg, I. (1996). *Mathematical Circles: (Russian Experience)*. Imprint Providence, R. I.: American Mathematical Society.

Frenkin, B. R. (2009). In B. R. Frenkin and N. N. Konstantinov (Eds.), *Letnie konferentsii Turnira gorodov* [*Summer Conferences of the Tournament of Towns*]. Moscow: MTsNMO.

Fuchs, D. B. (2007). Soviet mathematics in the 1950s and 1960s. In S. Zdraskovska and P. L. Duren (Eds.), *The Golden Years of Russian Mathematics* (pp. 212–222). Providence, R.I: AMS Publications.

Gelfand, I. M., Glagoleva, E. G., and Kirillov, A. A. (1973). *Metod koordinat* [*The Method of Coordinates*]. Moscow.

Kirillov, A. A. (1972). *Elementy teorii predstavlenii* [*Elements of Representation Theory*]. Moscow: Nauka.

Konstantinov, N. (2009). *Mezhdunarodnyj Turnir Gorodov* [*The International Tournament of Towns*]. Compiled by N. N. Konstantinov and C. A. Dorichenko. Moscow: Novoe Izdatel'skoe Delo.

Koval'dzhi, A. (Ed.). (2006). *Zapiski o vtoroj shkole.* [*Notes About School Number Two*]. Moscow.

Shifman, M. (Ed.). (2007). *You failed your math test, Comrade Einstein.* World Scientific.

Soifer, V. N. (1994). *Lysenko and the tragedy of Soviet Science.* New Brunswick, N.J.: Rutgers University Press.

Sossinsky, A. B. (2007). In the other direction. In S. Zdraskovska and P. L. Duren (Eds.), *The Golden Years of Russian Mathematics* (pp. 223–244). Providence, R.I.: AMS Publications.

Sossinsky, A. B. (1997). The Independent University in its new quarters. *Notices of the AMS, 44(4)*, 446–447.

Szpiro, G. (2007). Bella Abramovna Subbotovskaya and the "Jewish People's University." *Notices of the AMS*, 54(10), 1326–1330.

Taylor, P. J. (1992). *Tournament of Towns.* Belconnen, Australia: AICME Publ.

Taylor, P. J. (1993). *Tournament of Towns.* Belconnen, Australia: AICME Publ.

Taylor, P. J. (1996). *Tournament of Towns.* Belconnen, Australia: AICME Publ.

Taylor, P. J. (1998). *Tournament of Towns.* Belconnen, Australia: AICME Publ.

Tikhomirov, V. M. (2007). A. N. Kolmogorov. In S. Zdraskovska and P. L. Duren (Eds.), *The Golden Years of Russian Mathematics* (pp. 101–127). Providence, R.I.: AMS Publications.

Vasiliev, N. B. (Ed.). (2005–2006). *Zadachnik Kvanta, Matematika* [*Kvant's Problems, Mathematics*], Parts 1 and 2. Moscow: Nauka.
Vershik, A. M. (1990). Vladimir Abramovich Rokhlin 1919–1984 (on the 70th anniversary of his birth). *Algebra i Analiz* 2(2), 227–235. English translation: *Leningrad Math. J.* 2(2), 419–426 (1991).
Vyalyi, M. (1999). *Desyataya Letnyaya Konferentsiya Turnira Gorodov* [*Tenth Summer Conference of the Tournament of Towns*]. Moscow: MTsNMO.
Zvonkin, A. K. (2007). *Malyshi i matematika* [*Little Children and Mathematics*]. Moscow: MTsNMO.

6

Russian Traditions in Mathematics Education and Russian Mathematical Contests

Mark Saul

John Templeton Foundation, USA

Dmitri Fomin

Parametric Technology Group, USA

1 Introduction

Competitions in mathematics have a history stretching back to ancient times, and sometimes resulting in significant contributions to the field. Archimedes' Cattle Problem, whether or not it was meant as serious mathematics, was supposedly sent by Archimedes as a challenge to Eratosthenes and the mathematicians of Alexandria. The algebraists of the Italian Renaissance competed with each other in the solution of higher-degree algebraic equations, with results that have become classic. One of the earliest appearances of the result we now call Stokes' Theorem was on the Tripos competition at Cambridge, in Victorian times. Even today, the Clay Institute prizes can be seen as a large-scale competition in the solution of mathematical problems.

So it is no surprise that mathematical competitions arose independently in different venues during the 20th century. This chapter examines the traditions of Russian mathematics competitions in their cultural context, comparing them with developments in other countries, and particularly in the United States.

2 The Mathematical Subculture of 20th Century Russia

Like other very large countries, Russia is in many ways unlike its neighbors — if it can be said to have neighbors. Separated by a large land mass from the cultural centers of Western Europe, landlocked for much of its existence as a state, isolated at several times in history from outside influences, Russia is a part of Europe, yet culturally distinct. Its modern industrial economy was forced into existence, built on a quasi-feudal agrarian society and a burgeoning, but nascent, capitalism. For large parts of the 20th century, its scientific community was isolated by international politics and constrained by domestic policies.

Before 1917 most of Russia was agrarian and illiterate. Industrialization was taking place rapidly in the early years of the 20th century, but did not yet affect the bulk of the population. Cultural life was under strong political control. The revolutions of 1917, which eventually brought the Bolsheviks to power, were accompanied by severe difficulties, but also ushered in a decade full of hopes and dreams for many Russians. The evolution of new styles of art, literature, music, dance, and architecture, that had started in the last years of the Tsars, received new impetus and took unexpected turns, often supported by the newly formed Soviet state.

Then the totalitarian noose started tightening around the country's intellectual life. A mass emigration of talent depleted the artistic community. The Russian scientific community became isolated from the outside world. Young people with active minds saw only danger in a career in the arts. The sciences found themselves likewise constrained. Biologists and social scientists were quickly given limits to their research. Even the physical scientists, whose work might have immediate application both to industry and to war, found that their research efforts were controlled or challenged.

Mathematics, on the other hand, offered intellectual freedom (see Saul, 1992). Mathematicians needed no laboratories, and so the regime had no levers of control over their work. And the government needed this work, both ideologically and practically. Ideologically, because Communism ("Scientific Socialism") saw science and technology as building the future of humanity. (Indeed, despite ideological and political restrictions, science in general did well under the Soviets.) Practically, because the physical and natural sciences, which found direct industrial and military application, depended on the results of mathematicians for their more immediate fruitful results. So, in the planned economy, mathematics played the role of a heavy or extractive industry — supplying the intellectual base for further development.

On the positive side, the Soviet government needed mathematics and mathematicians for just this reason, and so encouraged the development of a mathematical community. Many mathematicians took seriously the need to industrialize the country, to modernize the Russian economy, and saw education as the key to the development of a vibrant mathematical and scientific community in Russia.

For all these reasons, mathematics attracted fine minds. And, because these minds had some unspoken, even unconscious anti-totalitarian agenda, the mathematical community began to assume the character of a subculture within Russian, or Soviet, society.

One of the characteristics of any such subculture is a need to "reproduce," to find new and younger members. More than in most mathematical communities, researchers took an active interest in education. Andrey Kolmogorov and Evgeny Dynkin were among a number of well-known mathematicians who founded actual high schools specializing in the study of mathematics. Israel Gelfand set up a "school by correspondence," reaching students in remote areas. It was common for graduate students, young faculty members, and even well-established researchers to return to their high schools or elementary schools and run "mathematical circles": clubs where younger students explored mathematics.

Much of this activity occurred in the 1960s, with Khrushchev's "thaw" and de-Stalinization. Many cultural historians have noted that scientific and artistic creativity is not directly tied to political freedom. Indeed, some of the world's greatest art and science was produced under strongly authoritarian political regimes. (Elizabethan England, Russia under the Romanovs of the mid-1800s, the various principalities of the Italian renaissance are examples). Often, a slight release of political pressure occasions an outpouring of creativity, and perhaps this is one reason for the peak in mathematical activity in Khrushchev's Russia. Another reason might be renewed opportunities to communicate with the outside world, to share and merge the knowledge developed in the international isolation imposed by Stalin.

3 The Form and Style of Russian Competitions

The Russian mathematical subculture produced a wealth of materials in mathematics, only now starting to be translated and made available in other countries. Competition, so natural to mathematics, assumed forms related to the cultural circumstances. For example, Russian competition rarely involved multiple choice or short-answer formats. Virtually all contest questions are open-ended, Olympiad-style. This is in marked contrast to American competitions, which started in short answer or multiple choice format, and developed Olympiad-style questions only in response to influence from abroad such as participation in the International Mathematical Olympiad.

One very simple reason for this tendency to avoid short answer questions was the lack of duplicating equipment. Mathematical competitions were often local affairs, organized by a few mathematicians in a single school, group of schools, or locality. The writers of the contests had official sanction, but no access to duplicating machines, mimeograph, dry copiers, or any technology that existed at the time. Such technology could easily be used for subversive activities, or to get around control of the media, so access to duplicating equipment was severely limited by the government.

Limits on printing or publishing have a long tradition in Russia. Readers of Dostoevsky's novel, *The Possessed* (*Бесы*), may recall that an illegal printing press is at the center of some of the action in that rather complicated plot. The writer himself was arrested for his association with a group that was branded as "revolutionary" by the government (see Grossman, 1975; or http://www.dartmouth.edu/~karamazo/

bio03.html). One of the charges against him was that he had planned to set up an illegal printing press. The technology available to the Soviet state only strengthened this tradition, which had unintended effects on education in general, and mathematics competition in particular. In Russia, and in the Soviet Union, testing in education was rarely by short answer. Test questions had to be written on a chalk board, or set orally, and so the given information had to be brief, yet stimulate thought on the part of the students. This form of question is nearly universal in Russian contests, on every level.

The tendency towards long answer, Olympiad-style problems was reinforced by the possibility of grading them. Because mathematicians and graduate students were available, and took an interest in nurturing the next generation, contest papers could be read closely and accurately by readers with considerable mathematical sophistication.

In the United States, this circumstance is difficult to arrange. The Tournament of the Towns, a Russian competition, routinely involves 50–80 papers read and graded for a large city. In Leningrad/St. Petersburg, it was not unusual to have 20–30 readers for such a competition. When the same competition was given in New York, only three readers were forthcoming (for 80 papers), despite an appeal to local universities and professional societies.

The Tournament of the Towns includes a summer program, the culmination of students' participation. Often, this includes an extended competition, in which students are given several days to think about and solve problems. These problems are posed orally, with some discussion about the problem situation between the posers (mathematicians) and the solvers (students).

Interaction between poser and solver is not uncommon in Russian mathematical competitions. Often, these situations are the first in which students meet working mathematicians, receiving inspiration for their work in a way that reaches adolescents directly and effectively. These situations also expose research mathematicians to students, keeping their interest in and skill with this population fresh and direct. Indeed, this sort of interaction can be viewed as one of the goals of the competition, which thus became much more than a process of selecting the brightest or quickest students.

Thus the historical and cultural background of Russian mathematics competitions influenced the form of the questions, the procedures for grading, and the interaction between grader and student. This background also influenced the actual content of the questions.

In general, there was an effort to make mathematics accessible to any student who was interested or motivated. This meant limiting the background necessary for solving contest problems to the mathematical knowledge provided by the school curriculum. The centralized nature of the Soviet educational system made it easy to determine the background that students could be expected to have at a given point in their education.

Happily, mathematics is endowed with an intellectual structure which allows very difficult and interesting questions to be asked with almost no background (Fermat's last theorem, Goldbach's conjecture, the four-color theorem, and so on). Luckily, the Russian school curriculum included arithmetic, pre-algebra, and geometry relatively early. By age 12, Russian students could be expected to understand the structure of complex word problems, of the type that are not given in America until students study algebra in high school. (Solutions for younger students were given in arithmetic terms only.) Because the study of geometry was "interleaved" with arithmetic and algebra (two or three days of one, two days of another, in separate classes), early introduction of geometric notions was also possible.

On the other hand, modular systems, often studied as an enrichment topic by American students, were generally avoided, both in stating problem and in writing solutions. Combinatorics and probability were not treated at all. Indeed, the traditional Russian curriculum, as the traditional American curriculum, was built around the need to prepare some students for calculus.

But perhaps the most significant influence on Russian contests was the effort to make mathematics available to all students, even those whose ability might be high but whose backgrounds were weak. This influence is often seen by Americans as "democratizing" or related to equity, and to American attempts to access the gifts of students in demographic groups underrepresented in the mathematical community. While the Soviet government had similar goals, this "leveling" tendency in the content of Russian mathematical contests had a slightly different origin.

Like most totalitarian regimes, the Soviets, concentrated power in the capitals of Moscow and St. Petersburg. In these cities were found the richest cultural resources, including universities, and the most active intellectual life. Again, this was a Russian historical tendency, exacerbated by the Soviets. Viewers of Chekhov's play *Three Sisters* may remember his characters' longing to go to Moscow, to the center of culture and society, and there are many references to this phenomenon throughout Russian literature. Students in provincial cities often had less access to university mathematicians, or to the best teachers, or to special schools.

There were, however, very notable exceptions to this rule, and to the Soviet attitude towards the distribution of cultural resources. Many mathematicians in the two capitals had their roots in other areas of the country and had come to Leningrad or Moscow in search of better education early in life. Often, they wanted to bring the fruits of their education back to the regions and small provincial towns from which they came. Since the infrastructure — telephones, roads, colleges — often was underdeveloped, and the distances often vast, they could only do that to a degree, so they had to select relatively large provincial centers as the nodes of the network and then proceed from there. Examples of local clusters of support within Russia proper include Novosibirsk, Tomsk, Kirov, Gorky (Nizhny Novgorod), Saratov, Arkhangelsk, Sverdlovsk (Ekaterinburg), and many others.

The same can be said about the ethnic minorities in the multicultural Soviet Union. Scientists coming to the capitals from smaller republics such as Armenia, Belorussia, Uzbekistan, or Latvia felt the need to return something to their home republics. In this case the flow of the mathematicians back to their native regions was even more strongly motivated, the attraction of the native culture more pronounced. It was much easier for a smart child from Gorky to come and stay in Moscow than for a smart child from Erevan, since all of his or her relatives and cultural connections would be back in Armenia, and the cultural bond was that much stronger.

The Soviet government too had an interest — or at least an expressed interest — in breaking down any nascent economic or social "classes" (in the Marxist sense). So there was agreement that children of workers and those of intellectuals should have the same opportunities for success in mathematical contests. The inclusion of problems requiring little or no background can be seen as part of this effort.

A third reason for the inclusion of problems requiring little background knowledge was the existence of a national curriculum. Competition gave the mathematical community a new way to identify talent. The regular school curriculum could be mastered by many, but one way to identify children who had still more potential was to give them problems to which they had access, but for which they were not specifically trained. Competitions were a perfect venue for this effort.

It is a common insight among writers of mathematical contests that the most difficult questions to compose are those that are interesting yet easy to solve, or that require little background. The harnessing of powerful mathematical minds to this task was, and continues to be, one of the significant achievements of the Russian mathematical community.

The samples of Russian mathematical contest questions given below illustrates some of the trends noted here.

4 The Growth of Russian Contests

4.1 *The Leningrad Mathematical Olympiad (LMO)*

Each Soviet republic, each big city, took great pride in its mathematical traditions, but the fiercest pride perhaps is that of Leningraders. Their city, once the capital of a great empire, and now eclipsed — administratively but not culturally — by Moscow, suffered with the disfavor of Stalin and also from a devastating siege during World War II.

The LMO was first organized in 1934, making it the oldest continuously held Olympiad in Russia (nosing out the Moscow Olympiad by just one year).[1] The year 1934 was a dark one in the Soviet Union. Stalin's Great Purges began that December with the assassination of his

[1] This account of the Leningrad Olympiads draws from two much more detailed accounts in Fomin (1994) and Rukshin (2000).

heir apparent, Sergei Kirov — who held the office of Leningrad party chair. One cannot help reading this coincidence as a turning inward of the mathematical community in the face of historical adversity.

The comments noted above about Russian contests are exemplified by some details of this first competition. Some of the most illustrious names in Russian mathematics appear in the list of the jury. A rule was made that a student winning in one year, no matter how young, could not participate the next year. The intention was to prevent domination of the contest by a small group of repeat winners. (This rule was later rescinded.)

Until very recently, the LMO had four levels:

(1) A school-based level. Contests are held in the schools for the top six grades. (The Russian system had 10 years of schooling.) Papers were graded by the schools' own mathematics teachers.

(2) A regional level, organized by geographical sections of the city. Theoretically, this was for winners in the school-level competition. But in fact almost any student was allowed to enter, even if they had not won in the first round. Typically, more than 10,000 students participated, in a city of less than 5 million.

(3) A citywide level, which was considered the main round. This round, remarkably, was oral, with about 100 students in each of grades 5–10 (as of 1990, grades 6–11) examined for up to 4 hours by research mathematicians. Problems were posed in preliminary "classes," in which the first four problems were given to the students. These problems were described orally and written on a chalk board or sheet of paper. Ambiguous or subtle points in the problem statement could be addressed by the posers. Students who solved some preset number of these first four problems (usually two, seldom three) were later given up to three more problems to work on.

The jury consisted of about 50 professors, university faculty, and graduate students, working in pairs. The grading was often a subtle and intellectually demanding process. Solutions were presented, after which students were examined, in the academic style, to probe the depth of their understanding of the problem and its solution.

(4) An elimination round, for students in the upper three grades. Up to 100 students took part in this second oral competition, which could last up to 5 hours.

Several points stand out, even from this brief description of the format of the contest. An oral examination of this length and depth could be carried out only with the full participation of the mathematical community, which was assured from the inception of the contest. Also, the point of the entire enterprise clearly was participation, with the winnowing out of prizewinners seen as a means to an end. The division, in the oral rounds, of problems was meant to ensure that as many students as possible succeeded in solving at least one problem. These students could be served efficiently. The jury could then concentrate on the smaller number of students who needed more challenging work. Rules of participation were worked out to allow multiple criteria for participation at various levels, and to account for different levels of preparation of the contestants.

A word about this last point is in order. Within the city, there were specialized schools for science and mathematics. There were also local "mathematical circles" which, among other goals, trained students for competition. Students inevitably made choices in their lives about what activities to participate in. Various provisions were made for students in special schools, so that their advantage against other students was minimized. Nonetheless, students from these schools typically shone in competitions.

The oral rounds offered a sort of accessibility to the students of the mathematical community. Rather than being a coldly official body, hiding behind a paper handed to students, the jury was alive in front of the candidates. Young students saw mathematicians caring about a mathematical problem, sometimes thinking about it in front of a group of students, and certainly taking pains to make the student understand the problem. Mathematics was presented to the student as something organic, alive, and developing, a center around which they could structure their lives (at a time when such structures were crumbling around them). Mathematical accomplishment could be seen as much as an effort by a community as by an individual in a competition — and this community was clearly interested in having young people join them.

Students also were able to experience first-hand the workings of a scientific community, with errors and fixes, formal proofs, the encouragement of questions or criticisms, and the participation of many people in the process. This was very different from the rather formal pedagogy they were exposed to in school.

4.2 *The Moscow Olympiad and the Development of the All-Union Olympiad*

One year after the first Leningrad Olympiad was held, a similar event was held in Moscow. A notable difference between the events is the absence in the latter of an oral round. Nonetheless, the participation of the mathematical research community was as important here as in Leningrad. The Moscow Olympiad both built on and contributed to the support this community gave to its prospective members in their pre-college years. For example, Tikhomirov (2006) writes that before the Olympiad there were few mathematical circles in Moscow (Tiknomirov notes that these institutions were pioneered by I. M. Gelfand, while still a graduate student). After the Olympiad, mathematical circles proliferated. The Moscow Olympiad continues in this role to the present.

The Leningrad and Moscow Olympiads, in the two most important cultural centers of the country, inspired mathematicians and educators in other cities to develop programs. The idea was adopted in intellectual centers such as Kiev,[2] Gorky (Nizhny Novgorod), and so on. By the late 1950s, such events were a local institution in many cities, and the Russian (or Soviet) mathematical community was ready to take the next step.

In a colorful account, Vasiliev and Egorov pinpoint this next step as occurring in Tbilisi, at a conference of topologists, in a tour bus on an excursion day. They describe[3] a discussion among several mathematicians, from different cities in the Soviet Union, who had the idea that winners of local Olympiads should participate in a contest on a higher level. (The writers mention B. N. Delaunay, I. V. Gersanov,

---

[2]An account of the history of the Kiev Olympiad can be found in V. A. Vyshensky *et al.* (1984).

[3]See Vasiliev and Egorov (1988). The present account of the growth of the national Olympiads summarizes the more detailed account given by these authors.

D. B. Fuchs, and A. S. Schwartz as participants in this seminal conversation.) The idea was brought back to Moscow, and, in 1960, winners of local competitions were invited to participate in the final round of the Moscow Olympiad. This was quickly recognized as a successful event, and the next year a similar event was organized, independent of the Moscow Olympiad, but held in Moscow at the same time. Although it was called the "All-Russian Olympiad," it sometimes included students from other Soviet Republics as well.[4]

This competition continued to be held in Moscow for some years, and added momentum to the rapidly growing culture of mathematical Olympiads, summer schools, and mathematical circles. One of the more interesting results was the so-called "Olympiads by Correspondence" (*Zaochniye Olympiadi*), organized in several areas of the country where distances made it difficult to have on-site events. An "All-Union Correspondence Olympiad" was started, appearing in national journals such as *Komsomolskaya Pravda*, the newspaper of the national Young Communist Organization (Vasiliev and Egorov, 1988, p. 10). The All-Russian Olympiad also tended to further cooperation among organizers of Olympiads in different regions, and particularly between the Moscow and Leningrad mathematical communities.

The success of the mathematical Olympiads on this larger scale also stimulated activity in other content areas. Olympiads in physics, biology, linguistics, and computer science developed in the years 1960–1980, and clearly received inspiration from the success of the mathematical community. A few experiments involved combining mathematics and physics, or mathematics and linguistics, and later mathematics and computer science, in a single event.

In 1966, the All-Russian Olympiad moved from Moscow to Voronezh. The contest had begun to outgrow both its venue and its name. Students from other Soviet republics (and not just the Russian republic) had been invited several times, and the identification with Moscow in particular and the Russian republic in general was weakening.

---

[4]The Soviet Union was made up of 15 Soviet semi-autonomous "republics," of which the Russian republic (RFSFR) was the largest.

Indeed, the next year saw the first "All-Union Olympiad," held in Tblisi (the capital of the Georgian Republic). This was a natural outgrowth of the "All-Russian Olympiad." Winners of Olympiads from throughout the Soviet Union were invited, as well as past Olympiad winners who were still students. In that year the Ministry of Education began to lend its support to these competitions, as well as those in physics and chemistry. The Olympiad had come of age.

During the ensuing decade, the All-Union Olympiad expanded and developed. Structures were put in place to articulate local and regional competitions with the national event. The format of the Olympiad was frequently a subject of discussion and investigation. An experiment in running an oral round, inspired by the Leningrad tradition, was reluctantly abandoned. Another experiment involved giving a lecture on the first day, then posing problems based on this lecture. Still other ideas involved a "research round," in which students would take their favorite contest problem and examine it in depth. Having gathered students from across the enormous country, the mathematical community wanted to use the occasion to its best advantage. The notion behind many of these experiments was to harness the motivation and energy generated by an Olympiad event to produce deeper, more sustained mathematical thought. In general, but with some exceptions, this step is now taken in other venues than the contest itself.

The advent of the International Mathematical Olympiad in 1967 was a great influence on the Soviet competitions. The format of the central event gradually became aligned with the two-day format of the international competition. Later developments also served to standardize the running of the competition. In 1975, the structure of the Olympiads was made more formal, with five levels distinguished: the school level, the city level, the regional level, the Soviet Republic level (i.e., each of the 15 republics), and finally the All-Union level (Vasiliev and Egorov, 1988, p. 17). This formal structure lent a more competitive nature to the upper levels. It also served to decrease the number of students participating in the All-Union round. Before this development, there were sometimes more than 500 participants in this Olympiad. The number of participants in later contests was less than 200. The mathematical community recognized this aspect of the institutionalization of their work as a mixed blessing. Nonetheless, the structure remains, and continues to inspire pre-college students in the development of their minds and their careers.

4.3 *"Math Battle"* (*Матбой*)

Few local Olympiads, and no national Olympiad, relied as heavily on oral examination as the LMO; however, several forms of competition were developed which did provide for interaction between students and mathematicians. Foremost among these was the "math battle" (матбой), a ritualized "combat" between two teams of students, mediated by a jury of mathematicians.

A math battle starts with the choice, by the jury, of a set of questions for the two teams to solve. The teams are then given a significant amount of time (several hours; one day; sometimes even a few days) to prepare solutions and strategy. Each team chooses a captain, and the two captains are given (orally) a short-answer question to solve. The team whose captain gets the answer quickest decides whether to go first or to give this right to the opponents.

If Team A goes first, it challenges Team B to the solution of one of the problems (not necessarily the first on the jury's list). A representative from Team B gives a solution, and someone from Team A tries to find flaws in the solution. Team B must respond to these criticisms. The jury can also critique the solution. If Team B cannot answer an objection raised by Team A, then Team A becomes the "solving" team if they so wish. They can respond to their own objection and even construct a new proof and earn the points, with Team B now trying to find holes in Team A's reasoning.

The proceedings are oral, with a chalk board constantly in use, and with both teams, the jury, and sometimes the public as audience.

When everyone agrees, discussion of this first question concludes. The jury then deliberates, awarding points to each team, and sometimes to itself, for correct solutions or meaningful criticisms.

It is then the turn of Team B to challenge Team A to the solution of a problem. The process is repeated with the teams exchanging roles. This process continues until the list of problems is exhausted, or until neither team has a solution to any more of the problems.

The situation is complicated by a rule allowing for "reversal of roles." That is, if Team B challenges Team A to solve a particular problem, Team A can choose not to accept the challenge. In this case, Team B must offer a solution, and Team A must critique the solution. That is, the roles are reversed: Team B is solving the problem, although it is this team which originally issued the challenge. At the conclusion of Team B's solution (to their own challenge problem), it is the turn of Team A to offer a challenge. If Team B accepts this new challenge, it will find itself in the position of offering a solution for a second time in a row. Or, it can reject the challenge, and the roles reverse once again.

Math battles are typically very attractive to students. Tales are told of teams staying long beyond the closing of the Metro, so students must walk home in the dark and the cold. Elaborate local rules were often worked out, to cover special cases or infractions that came up.

Perhaps the most noteworthy aspect of a math battle is the engagement of the jury on a direct and equal intellectual level with the contestants. Students have the opportunity to engage in meaningful mathematical dialogue with working researchers and graduate students.

4.4 *Tournament of the Towns*

An interesting outgrowth of Russian mathematics competitions was the establishment of an international contest, based on Russian models. The Tournament of the Towns was organized in 1980 by a group of Moscow mathematicians who had long been engaged in the selection of Olympiad winners on a national level. Among them, N. N. Konstantinov stands out as having played a major role (see http://www.turgor.ru/, or http://www.amt.edu.au/imtot.html).

The contest is given twice a year, in two levels: Russian grade 10 and Russian grade 11. Local organizers throughout the world choose which level to give to which students. They also choose a day (within a given range) to administer the competition and make local arrangements. The competition is written, rather than oral. Papers are graded initially by a local team, then sent to Moscow for final judging. Prizes are awarded both locally (i.e., in each city) and internationally.

As the name indicates, competition is between cities or municipalities, not countries or teams. Each city is represented by one paper for each 100,000 inhabitants, with a minimum entry of five papers. Eligibility and selection of team members is all defined locally.

Participation in the Tournament of the Towns has spread knowledge of Russian competition traditions, and of the high level of mathematics possible within a pre-college competition, to areas of the world which have not yet developed their own local traditions and institutions.

Winners of the competition receive a certificate of achievement, and also an invitation to a summer program, which is not always held in Russia. At this summer program students engage in a math battle, listen to lectures, and interact with the organizers, who are themselves sometimes outstanding mathematicians.

The Tournament of the Towns summer program is part of a network of such programs in mathematics, often of great importance to the development of students' mathematical interests, but connected to the system of competitions only in that the two institutions serve many of the same students.

4.5 *Kvant: A Competition in a Journal*

Among the most significant institutions serving high-achieving students in the Soviet Union was the journal *Kvant*. Founded in 1970 by a group of scientists (notably including Kolmogorov and Kikoin), it made high-level content in mathematics and physics available to pre-college students, including those who did not live in large cities. Indeed, one of its functions was to fill in the "gap" between such places and the cultural centers, by forming a virtual community of problem solvers long before the internet made such communities commonplace.

Along with the articles, which were themselves interactive, with problems posed alongside the exposition, there was a problem-solving column. Readers were invited to submit solutions, with lists of winners and books as prizes. This "virtual competition" articulated with other competitions in that winners of the *Kvant* "contest" were often invited to participate, on the basis of this achievement, in local olympiads.

The journal *Kvant* was particularly meaningful for students who did not live near a local specialized mathematics and science school, or could not find a nearby mathematical circle where they could meet other students with the same interests and abilities. It was extremely popular in the 1970s and 1980s.

5 What Can We Learn?

The intent of the outline given above of Russian mathematics contests is to set this phenomenon in the context of a mathematical community. This community, in turn, is deeply affected by its own historical, political, and geographic context.

So it is not simple to extract a lesson from the Russian experience, in the sense of advice to another mathematical community, in another context. We can say, however, that the success of the Russian system of competitions is due in large measure to the involvement of the entire mathematical community: students, teachers, graduate students, and research mathematicians, who saw their career as a continuous movement through a system of support.

In many countries, the education community and the research mathematics community are kept apart by the parochial interests of their institutions. In some cases, the two communities are at odds, with differing understandings of each other's role. While cases of this sort of friction certainly can be found in the history of Russian mathematics, they are overwhelmed by instances of cooperation, forged by historical and political circumstances.

It is not likely, and probably not desirable, to reproduce in other country the circumstances that led to the Russian system of mathematical competitions. But the view that they are a reflection of the efforts of an entire mathematical community is one from which other communities can learn.

6 Sample Problems

1. The integers 1 through 64 are written on an 8×8 chessboard, one on each square. Show that there exist two neighboring squares

with numbers which differ by 5 or more. (Two squares are neighbors when they share a common side.)

Solution: Suppose the assertion were false. Let us choose the numbers 1 and 64, and proceed from one to the other through a shortest path consisting of neighboring squares, adding up the differences between the numbers we encounter. Since each difference is 4 or less, the sum of these differences is at most $4(k-1)$, where k is the number of squares in our path. The path cannot consist of more than 15 squares, and therefore this sum is less than or equal to $4 \times 14 = 56 < 63 = 64 - 1$. This contradiction proves that the original assertion must be true.

Source: Moscow Olympiad, 1963, second round, 8th grade.

Note: This problem involves logic and arithmetic, but no other mathematical background. The proof by contradiction given above is a sophisticated one for students of this age, who usually give a cumbersome "proof" with many cases. Nonetheless, younger students can profit by experimentation with the numbers, and often come away from the experience with an appreciation of arguments by contradiction.

2. Can the numbers 1, 2, 3 . . . 13 be placed around a circle so that any two adjacent numbers differ by 3, 4, or 5?

Solution: This is impossible. No two elements of the subset $S = \{1, 2, 3, 11, 12, 13\}$ can be adjacent, and there are seven remaining numbers which can separate them. Since (in a circle) the six elements of S determine six 'spaces', it follows that one pair of elements of S must be separated by two numbers, and the others by only one number.

Of the elements of S, only the number 1 can be adjacent to 4. Hence the number 4 cannot be placed between two elements of S. Similarly, of the elements of S, only the number 13 can be adjacent to 10. Hence the number 10 cannot be placed between two elements of S. It follows that 4 and 10 must be placed next to each other (in the 'space' holding two numbers), which contradicts the requirements of the problem.

Source: First All-Union Olympiad, 1967, grade 10.

Note: Students often find this an easy problem. The argument emerges quickly out of experimentation. Their struggle is usually to find wording for the proof.

3. For many years, Baron Munchausen has been going duck shooting on a lake every day. Starting on August 1, 1991, he tells his cook each day: "Today I shot more ducks than two days ago, but fewer ducks than a week ago." What is the largest number of (consecutive) days on which the Baron can say this? (Remember that Baron Munchausen never lies.)

Answer: 6 days.

Solution: Let us start following the Baron's story on July 30. We will trace his hunting record, with the number of ducks he shot increasing each day we mention.

According to his statement, he shot more ducks on August 1 than on July 30, and still more on August 3, more on August 5, and more on August 7. He shot more ducks on July 31 than on August 7, because that was "a week ago". Then he shot still more ducks on August 2, more on August 4 and more on August 6. But July 30 is a week before August 6, so he shot more ducks on July 30 than on August 6. Tracing the history backwards, this means that he shot more ducks on July 30 than on August 1, and therefore more ducks on July 30 than on July 30!

This contradiction shows that if he made his statement on August 1, he could not have made it (truthfully!) on August 7. The diagram below summarizes the argument (arrow $A \to B$ means that Baron shot more ducks on day B than on day A):

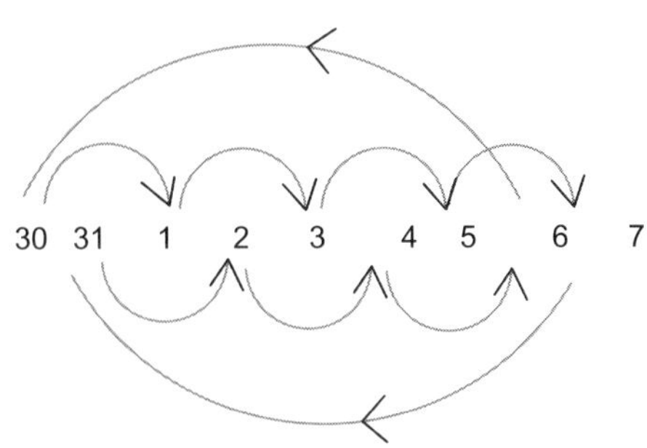

However, he can make the statement six days in a row, without contradiction. An argument can be constructed for this using the diagram below:

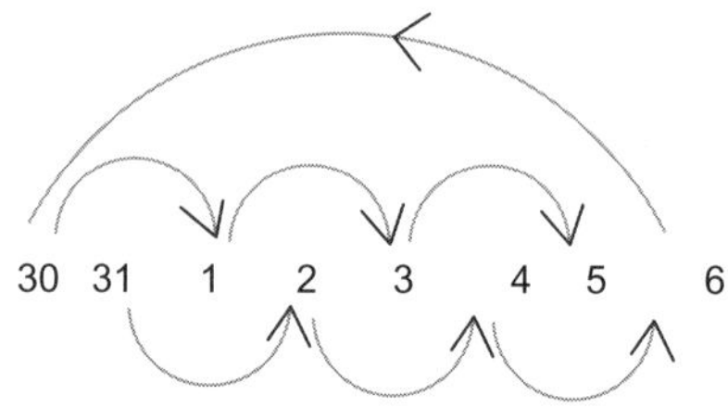

Note: This problem requires little mathematical background. Students are often able to draw a diagram such as we have given, but have trouble writing words to match. In judging a mathematical competition, care must be taken to uniformly value (or deny value to) such a solution presented graphically.

Russian students are quite familiar with Baron Munchausen, the legendary braggart and prevaricator, who appears regularly in Russian contest problems. Often (although not here) his name is a clue that the solver should prove that the situation described is impossible.

Source: Leningrad Mathematical Olympiad, 1991, third round, grade 6.

4. Is it true that, of any set of 100 different integers, we can choose a subset of (a) 15 (b) 16 integers such that the difference between any two elements of the subset is a multiple of 7?

Answers: (a) yes; (b) no.

Solution: (a) The difference of two integers is a multiple of 7 if and only if the two integers have the same remainder upon division by 7. There are only seven such remainders: 0, 1, 2, 3, 4, 5, 6.

Suppose that it were impossible to choose a subset of 15 integers such as is required. That would mean that no more than 14 of the original integers were themselves divisible by 7, no more than 14 had a remainder of 1 upon division by 7, no more than 14 had a remainder of 2, and so on. But then there could be no more than $14 \times 7 = 98$ integers in the original set. This contradiction shows that our assumption is incorrect, and there must be a set of 15 integers satisfying the conditions of the problem.

Note: This solution is a typical argument using the "pigeon hole principle": *If we have $nk + 1$ pigeons to assign to n pigeonholes, then in one pigeon hole there must be at least* $k + 1$ *pigeons*. Indeed, if there were no more than k pigeons in each pigeon hole, then there would be no more than nk pigeons altogether, a contradiction.

In Russian this statement is called *Dirichlet's principle*, and the canonical animals used to describe it are rabbits (in cages), not pigeons.

(b) We give a counter example. Take the set of integers from 1 to 100. There are 14 integers which have a remainder of 0: 7, 14, ...98. There are 15 integers which have a remainder of 1, and 15 with a remainder of 2. For remainders 3, 4, 5, 6, there are 14 numbers each in the original set. Therefore, there is no subset of 16 integers with the same remainder when divided by 7, and thus no subset of 16 with the required property.

Note: We have presented the solution as given in the original Russian source. In American texts, this would probably have been described using arithmetic modulo 7. Modular arithmetic is not introduced in Russian curricula, and so contest problems are stated and solved without using this tool.

Source: Problem 1–12, from Vasiliev *et al.* (1981) (the problem is on page 5, solution on page 15). This is a book of problems from the "Olympiad by Correspondence" run by Moscow State University.

5. A circle G and a point K are given on the same plane. Through two arbitrary points P, Q of the circle, and through K, a second circle is drawn. Let M be the intersection of line PQ with the tangent to this new circle at K. What is the set of possible positions for M?

Solution: Let O be the center of circle G, and let r be its radius. We draw tangent $\overleftrightarrow{MN}$ to circle G (see diagram). Now the square of a tangent from a point outside a circle is equal to the product of any secant through that point and the secant's external segment (prove this!), so we have:

$$|MK|^2 = |MQ| \times |MP| = |MN|^2,$$

so

$$|OM|^2 - |MK|^2 = |OM|^2 - |MN|^2 = |ON|^2 = r^2.$$

It is not hard to show, for example by using the Pythagorean theorem, that M must lie on a certain line l, perpendicular to OK, which is the set of points such that the difference of the squares of their distances to O, K is the constant r^2. (See, for example, the book by Gutenmakher and Vasiliev, published in English in 2004.)

Conversely, we can prove that any point on line l belongs to our set. Given a point M on this line, we can draw any circle tangent to line MK at K which intersects G at two points P, Q. Then line PQ will intersect line l at M.

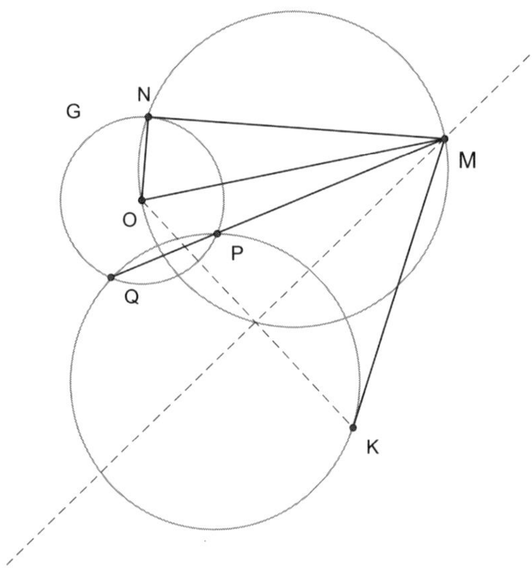

Note: This problem, like many others, draws on the rich literature available in Russian about problem solving, or in the tradition of elementary mathematics from an advanced standpoint.

Source: *Kvant*, M630 (by I. Sharygin).

6. The sides of triangle have lengths a, b, c, and $m\angle A = 60°$. Show that

$$\frac{3}{a+b+c} = \frac{1}{a+b} + \frac{1}{a+c}.$$

Solution: The given equation is equivalent to:

$$\begin{aligned} 3(a+b)(a+c) &= (a+b+c)(2a+b+c), \\ 3a^2+3ac+3ab+3bc &= (a+b+c)^2+a(a+b+c), \\ a^2+bc &= b^2+c^2. \end{aligned}$$

Thus the required relationship is equivalent to $a^2 = b^2 + c^2 - bc$. But we can obtain this relationship from the given information by applying the law of cosines:

$$\cos A = \cos 60^\circ = 1/2, \text{ and } a^2 = b^2 + c^2 - 2bc \cos A.$$

Source: 14th Belorussian SSR Olympiad, 1964, problem 11.1, see http://www.problems.ru/view_problem_details_new.php?id=109006.

Note: This is a problem in algebra, not geometry, although its motivation is geometric. The mention of an angle of 60° is a hint to use trigonometry, or at least the properties of special triangles. Students can in fact derive the special case mentioned of the law of cosines without knowing the general result, although such a derivation is rare. The mixture of algebra and geometry (and even trigonometry) is unusual in classroom work, but almost routine in competition settings.

7. A circle is covered by a number of arcs. These arcs may overlap, but no single arc covers the entire circle. Show that we can choose a subset of these arcs whose (central) angles sum to at most 720 degrees, but which still cover the circle completely.

Solution: If three of our arcs cover some part of the circle, and they all have some part in common, then we can choose two of these arcs which cover the same part of the circle. Let us eliminate the unnecessary arc in each such case (consecutively, not all at the same time). We will then have a covering of the circle in which each part of the circle is covered at most twice. This means that the sum of the central angles of the remaining arcs is at most 720 degrees.

Source: 15th Belorussian SSR Olympiad, 1965, problem 10.3, see http://www.problems.ru/view_problem_details_new.php?id=109006.

Note: This problem is difficult to classify. It is related to, but not typical of, combinatorial geometry. The notion of "covering" is important in measure theory, but the situation described here is atypical of that branch of mathematics as well. Part of its appeal as a competition problem is exactly that it resists classification.

8. A set of k points are given on the plane, such that every line containing two of the points will also contain a third point of the set. Show that all k points lie on the same line.

Solution: Suppose that not all these points lie on the same line, and consider the set of lines passing through at least two of the given points. We look at all the distances from the given points to each of these lines. We have assumed that these distances are not all 0. Suppose the smallest non-zero distance is between some point D and some line m, which contains points A, B, and C (see diagram). We draw altitude $\overline{DH}$, perpendicular to m, of triangle ADC. At least two of the given points must lie on one side of point H along line m. (We consider point H itself as belonging to both 'sides' of line m.) Without loss of generality, we assume that these are points A and B (as in the diagram). We can show that the distance from B to line $\overleftrightarrow{AD}$ is strictly less than DH. Indeed, the distance from H to line $\overleftrightarrow{AD}$ is strictly less than DH, since the altitude of a triangle is shorter than the sides adjacent to the altitude. And the distance from B to $\overleftrightarrow{AD}$ is clearly not greater than the distance from H to $\overleftrightarrow{AD}$. Thus the distance we thought was minimal is in fact not minimal. This contradiction shows that all the given points must lie on one line.

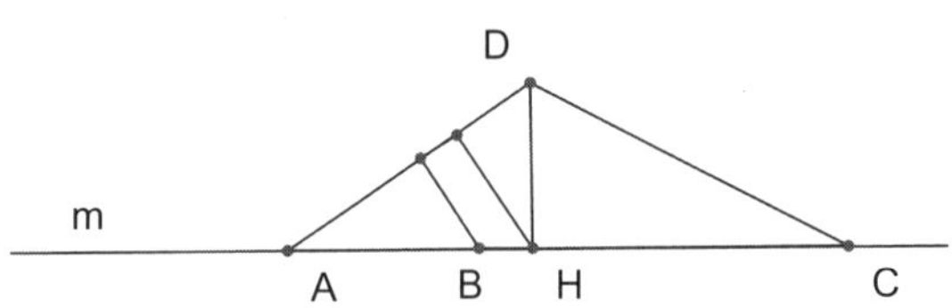

Source: 14th Belorussian SSR Olympiad, 1964, problem 11.2, see http://www.problems.ru/view_problem_details_new.php?id=109006.

Note: This result is typical of combinatorial geometry, a field which has produced many interesting elementary problems. Russian contest problems often draw from this source.

9. Let h be the smallest altitude of a given tetrahedron, and let d be the smallest distance between two of its opposite edges. For what vales of t is the inequality $d > th$ possible?

Answer: The inequality is possible for $t < 3/2$.

Solution: Let ABC be the face of tetrahedron $ABCD$ which has the largest area. Then the volume of the tetrahedron $V = \frac{S_{ABC}h}{3}$. On the other hand, this volume is equal to one sixth of the product of the lengths of two opposite edges, the distance between them and the sine of the angle between them. It is helpful to introduce a new tetrahedron $A'B'C'D$, where $A'B'C'$ is the triangle obtained by taking the vertices of ABC as midpoints of $\overline{A'B'}$, $\overline{B'C'}$, $\overline{C'A'}$. Then, for example, the area of $A'B'D = AB \cdot CD \sin f$, where f is the angle between AB and CD. Since the sum of the areas of three faces of a tetrahedron is greater than the area of the fourth face, the area of $A'B'C'$ is not greater than triple the largest area from among triangles $A'B'D$, $B'C'D$, $C'A'D$. This largest area corresponds to the minimal value of the distance between edges. Suppose this distance is $AB \cdot CD \sin f$. Then, using the fact that the area of ABC is one fourth of the area of $A'B'C'$, we obtain $V = \frac{1}{6}dAB \cdot CD \sin f = \frac{1}{3}h \cdot S_{ABC} \leq \frac{1}{3}h \cdot 3 \cdot \frac{1}{4}AB \cdot CD \sin f$ that is, $d < h/2$.

This estimate cannot be improved: if we take a regular pyramid, and let its altitude tend toward 0, the ratio d/h will tend toward $3/2$.

Source: All-Russian Mathematics Olympiad, 2008, problem 24.

Note: Solid geometry was part of the national curriculum, and contest problems often involved this content area. Inequalities also frequently appear in Russian contests. Many results can be obtained which are not easily intuited, but rely on the simplest of inequality results. This example involves a deeper knowledge of metric results in space geometry than most inequality problems.

10. In a senate consisting of 30 senators, each pair of senators is either friends or enemies, and each senator has exactly 6 enemies. How many triples of senators are there, in which each pair of the three senators are friends or each pair are enemies?

Solution: Let x be the number of triples described in the problem, and let y be the number of remaining triples. Then $x+y = \frac{30 \cdot 29 \cdot 28}{1 \cdot 2 \cdot 3} = 4060$. Suppose each senator gives us a list of triples he or she belongs to, such that the other two senators are either both his friends or both his enemies. Each such list will have $\frac{23\cdot22}{2}+\frac{6\cdot5}{2}=268$ elements. If we put together the 30 such lists, there will be $30\cdot268=8040$ triples listed altogether.

Note that this long list contains each triple. First, it certainly contains the triples we are after, and it contains each such triple 3 times (once for each participating senator). And each triple we are not interested in will be listed exactly once. Indeed, such a triple consists either of two friends who are both enemies of the third senator, or two enemies, both of whom are friends of the third senator. In either case, this triple will be listed by, and only by, the 'odd' senator.

Thus we have $3x+y=8040$, and $x+y=4060$. Solving simultaneously, we have $x=1990$.

Source: *Kvant*, M1244 (D. Fomin).

Note: This exercise in combinatorics is typical of Russian contest problems in this area. The problems are often difficult, as this one is, yet use only the simplest of tools. Solvers of this problem need only to know how to count combinations of two and three elements of a set, and not the general binomial coefficient. There are several other techniques here that generalize. The technique of making an inclusive list, then examining it for repetitions, is a useful one in combinatoric problems, but it is not easy to discern this path in the present solution. Behind several of the steps in the solution is the notion of partitioning a set (here, a set of triples). The use of two variables is in some ways an artifact of describing the solution, yet it would be difficult to express the solution without this unusual device.

11. A pack of 2002 cards with the numbers $1, 2, 3, \ldots, 2002$ written on them are put on a table face up. Two players in turns pick up a card from the table until all cards are gone. The player for whom the last digit of the sum of all the numbers on his cards is larger than his opponent's, wins. Who has a winning strategy and how should one play to win?

Solution: The first player wins. Let us pair up all the cards (numbers), pairing k with $1000+k$, $k=1,\ldots,1000$. We also pair 2001 with 2002. So in each pair except the last one both cards have the same last digit.

The first player starts by picking up 2002. From this moment on his strategy is to pick up the other half of the pair chosen by the second player. So, eventually the second player is forced to pick up 2001. More generally, if the cards are not gone, and all cards that are picked up constitute pairs, then the first player takes any card, say k'. Then the second player will eventually have to pick up the card with which k' is paired. Indeed, every time the second player selects some other card, then the first player selects the card with which this other one is paired. So, eventually, the second player will have to take the card paired with k'. At the end of this process, the first player has a sum that has the same last three digits as $1+2+\cdots+1000+2002 = 502502$, which has a final digit of 2, while the second player has a sum whose final digit is 1.

Source: Tournament of the Towns, Junior O level, fall 2002, see: http://www.math.toronto.edu/oz/turgor/archives.php.

Note: The argument is essentially one from symmetry. Some of the most interesting Russian competition problems are couched in the setting of a game. Even younger students can play the games, in this case using a much smaller set of numbers. Often, they can find a win through the play itself, although they have a difficult time expressing the solution — or even seeing the winning strategy as a "solution." For much more on this topic, see Fomin, Genkin, and Ittenberg (1993).

12. We are given a connected graph with n edges. Show that we can number the edges with the numbers 1 through n so that for any vertex through which two or more edges pass, the greatest common divisor of the numbers assigned to these edges is 1.

Solution: We choose some vertex V_0, and follow a path along the edges of the graph, assigning them the numbers $1, 2, \ldots, s$, until this is impossible.

If this process assigns numbers to all the edges, we are done. Indeed, if we arrived at some vertex by following an edge numbered k, we left that vertex at edge $k+1$. The greatest common divisor of these two numbers is 1, so the greatest common divisor of all the edges ending at that vertex is also 1.

If there are edges left after one round of this process, then the fact that the graph is connected assures us that at least one of the remaining edges goes through a vertex we have already visited. We start at that vertex, and again travel along the unnumbered edges, assigning them the numbers $s+1, s+2, \ldots$, until this is impossible.

Following this plan, we will eventually have assigned numbers to each edge of the graph. We will show that this assignment fulfills the requirements of the problem. Indeed, let us examine some vertex V of the graph. If this is V_0 (the initial vertex), then one of the edges ending at this vertex bears the number 1, which must then be the greatest common divisor of the numbers on all the edges at that vertex.

If V is not V_0, then suppose r is the smallest number attached to any edge ending at V. Then we must have arrived at V for the first time at the rth step in our process. But at that point, there was some unnumbered edge leaving V, to which we gave the number $r+1$. The greatest common divisor of r and $r+1$ is 1, which is therefore the greatest common divisor of the numbers assigned to all the edges ending at V.

Source: *Kvant*, M1318.

Note: Graph theory is yet another area of mathematics with many difficult yet accessible problems. Students need no formal instruction in this field to work problems which can lead to significant results.

13. Does there exist a convex polyhedron such that any cross section of the figure formed by a plane not passing through a vertex is a polygon with oddly many sides?

Hint: Move the cutting plane parallel to itself, and examine the change in the parity of the number of edges it intersects as the plane passes through a vertex.

Answer: There is no such polyhedron.

Solution: Suppose such a polyhedron existed. Then it must have oddly many edges ending at each vertex. Indeed, if some vertex A of the polyhedron were the endpoint of an even number m of edges, then we could form a cross section of the figure with m sides by cutting it with a plane which separates vertex A from its other vertices.

Now we consider a plane p such that any plane parallel to p contains at most one vertex of the polyhedron, and examine the family of cross sections of the polyhedron which are in planes parallel to p. Suppose we move p so that it remains parallel to itself. The number of sides of a cross section will change only when the plane passes through a vertex, and we have selected p so that the plane passes through only one vertex at a time. Suppose the plane has come to some vertex at which k edges end (where k, as we have seen, must be odd). If, before passing through this vertex, the plane intersected s of these k edges (where s, by hypothesis, is odd), then after it passes through the vertex it will intersect $s-k$ edges. Therefore, the number of edges intersecting the plane changes by $s - 2k$ (where $s - 2k$ is odd). This means that if the cross section was a polygon of n sides before the plane crossed the vertex, then after crossing the vertex, the cross-section is a polygon of $n + s - 2k$ sides. But n and $(n + s - 2k)$ are of opposite parity, so one of them must be even, contrary to hypothesis. This contradiction shows that the required polyhedron cannot exist.

Source: *Kvant*, problem 903, 1985. Author: A. Dorogovtsev.

Note: The original problem also asked whether there is a corresponding polyhedron with only "even" cross sections. An analysis similar to the one above will show that this property is possessed by any polyhedron with evenly many edges ending at each vertex (an example is a regular octahedron).

This problem is typical of those in combinatorial geometry, which frequently appear in Russian Olympiads. The argument barely uses any combinatorics at all, and includes the typical device of moving a plane parallel to itself and examining the changes in the situation being studied.

The art of giving hints to students is a difficult one to master. Hints are only rarely given in Olympiad situations.

References

Fedorov, R. M., Kanel-Belov, A. Ya., Koval'dzhi, A. K., and Yashchenko I. V. (2006). *Moskovskie Matematicheskie Olimpiady, 1993–2005*. [*Moscow Mathematical Olympiads, 1993–2005*]. Moscow: Izdatel'stvo MTsIMO.

Fomin, D., Genkin, S., and Itenberg, I. (1993). *Mathematical Circles: Russian Experience*. Providence, R.I.: American Mathematical Society.

Fomin, D. V., and Kirichenko, A. (1994). *Leningrad Mathematical Olympiads, 1987–1991*. Chelmsford, MA: Mathpro Press.

Fomin, D. V. (1994). *Sankt-Peterburgskie Matematicheskiye Olimpiady* [*Saint Petersburg Mathematical Olympiads*]. St. Petersburg: Izdatel'stvo Polytekhnika.

Grossman, L. (1974). *Dostoevsky: A Biography*, translated by Mary Mackler. London: Allen Lane.

Gutenmakher, V. L., and Vasiliev, N. B. (2004). *Lines and Curves.* Boston: Birkhauser.

Rukshin, S. E. (2000). *Matematicheskie Sorevnovaniya v Leningrade-Sankt-Peterburge* [*Mathematics Competitions in Leningrad/St. Petersburg*]. Rostov-na-Donu: Izdatel'stvo Tsentr "Mark."

Saul, M. (1992). Love Among the Ruins: The Education of High-Ability Mathematics Students in the USSR. *Focus, 12(1)*, 1–2.

Tikhomirov, V. M. (2006). Razmyshleniya o Moskovskikh Matematicheskikh Olimpiadakh. [Reflections on the Moscow Mathematical Olympiads.] In R. M. Fedorov, A. Ya. Kanel-Belov, A. K. Koval'dzhi, and I. V. Yashchenko, *Moskovskie Matematicheskie Olimpiady, 1993–2005* (pp. 7–15). Moscow: Izdatel'stvo MTsIMO.

Vasiliev, N. B., and Egorov, A. A. (1988). *Zadachi Vsesoiuznykh Matematicheskikh Olimpiad* [*Problems from the All-Union Mathematical Olympiads*]. Moscow: Nauka.

Vasiliev, N. B., Gutenmacher, V. L., Rabbot, Zh. M., and Toom, A. L. (1981). *Zaochnye Mathematiticheskiye Olimipiady* [*Mathematical Olympiads by Correspondence*]. Moscow: Nauka.

Vyshensky, V. A., Kartashov, N. V., Mikhailovsky, V. I., and Yadrenko, M. I. (1984). *Sbornik Zadach Kievskikh Matematicheskikh Olimpiad* [*Collected Problems from the Kiev Mathematical Olympiads*]. Kiev: Vischa Shkola.

7

The Relevance of Russian Elementary Mathematics Education

Jean Schmittau

State University of New York
Binghamton, NY, USA

1 Introduction

The major thesis of this monograph is that the relevance of Russian elementary education resides in its manner of addressing important issues that continue to be intractable or remain incompletely resolved in mathematics education. These include such issues as the role of so-called "real world" problems, formal (academic) vs. informal (market and workplace) mathematics, conceptual understanding vs. procedural competence, the role of calculators in the elementary school, early algebraic development, children's understanding using manipulatives vs. their failure to understand at the symbolic level, and the teaching of mathematics as a connected body of knowledge.

Many of these issues have arisen and their solutions have been approached within the framework of formal logic, that tends to isolate concepts and perceive their existence in relative isolation or within apparent dichotomies. By way of contrast, the resolution of these and similar issues in the Russian elementary program of V. V. Davydov is the result of the application of Vygotskian theory and dialectical logic in the construction of the curriculum. In addition to the above issues a number of apparent dichotomies are also resolved within the Davydov curriculum, including: equality vs. inequality, discrete quantity vs. continuous quantity with the attendant counting vs. measurement distinction, problem solving vs. practice on routine exercises, action on objects vs. action on symbolic forms, and mental calculation vs. calculator usage. Further, fundamental mathematical *actions* such as addition and subtraction, and multiplication and division, rather than being perceived as separate operations are viewed in dialectical complementarity. Other distinctions are resolved, such as those between the numerical and the algebraic, the abstract and the concrete, concepts and procedures, real world mathematics and decontextualized mathematics, instructional content and instructional methodology. The Davydov curriculum is not, of course, the only Russian elementary mathematics curriculum, but it is the one the author finds of greatest interest in the manner of resolving of the issues indicated above.

It will be argued that Vygotskian theory and dialectical logic function to resolve such issues by connecting apparently disparate concepts through their origins in activity, and as a consequence such concepts arise in their dialectical complementarity rather than as separate entities within a formal logical development. In this way the connections necessary to understand mathematics as a conceptual system are developed and reinforced throughout the elementary curriculum.

2 Russian Vygotskian-Based Mathematics Education Innovations

V. V. Davydov and his colleagues sought to eliminate the gap between elementary and secondary mathematics study, noting that the first focused on number and arithmetic and the second on the theoretical understandings of algebra. The relevant distinction between these, namely that of empirical vs. theoretical learning, was identified by Vygotsky (1986), who characterized empirical learning as resulting from children's ordinary interactions within the natural and cultural environment, and theoretical learning as occurring largely in academic settings and rendered accessible to students through the mediation provided in such settings. Unlike empirical learning, theoretical (or scientific) understanding requires probing beneath the surface characteristics of phenomena. An apt example is that of the diurnal cycle, the empirical view of which is linguistically conveyed by phrases such as the sun "rising" in the morning and "setting" in the evening, whereas the theoretical or scientific understanding of this phenomenon is that of the rotation of the earth on its axis (Lektorsky, 1984; Kozulin, 1990).

Within an academic context an empirical approach to learning the concept of circle, for example, might proceed inductively by asking children to abstract the concept from objects such as round plates or discs, wheels, etc. (Davydov, 1990). Although similar inductive practices may find acceptance within U.S. classrooms, it is questionable whether individual children's "constructions" would conform to the culturally constructed understanding of circularity that has survived the required epistemological testing within the academic field of mathematics. Such constructions, in fact, may reflect only what Hegel termed a "formally general" rather than the "fully universal" concept of circularity as defined by the mathematics community. A theoretical approach, however, such as that advocated by Spinoza and endorsed by Vygotsky, would ask children to fix one end of a string and rotate the other, thereby producing a circle and abstracting its essence as a path which is everywhere equidistant from a fixed point. This understanding not only would accord with the culturally accepted concept, but would enrich subsequently their experience of all objects and phenomena that exhibit circularity, and is an apt illustration of the ascent from the abstract to the concrete advocated by Hegel.

Vygotsky's thesis that each age in a child's development is characterized by a particular form of social activity led to further research by D. B. Elkonin (1975), who contended that the elementary school years were the most advantageous for the introduction of theoretical learning. Hegel (1956, cited in Davydov, 1990) also had argued that children should not be kept for long in an empirical mode of learning and, consequently, Davydov sought to introduce children to theoretical thinking in mathematics in the elementary grades in marked contrast to the Piagetian imperative to delay learning that required abstract reasoning until adolescence.

In further contrast to his U.S. counterparts, Davydov also rejected set theory as a basis for mathematics (Davydov, 1975a). Rather, he agreed with Bourbaki (1963, cited in Davydov, 1975a), that the essence of mathematics is mathematical structure which, in the case of the real numbers, is comprised of both order and algebraic structures. It was by no means obvious, however, how young children entering school could begin the study of algebraic structure, since they lacked both the requisite mathematics background and the advanced perspectives of mathematicians. Davydov (1975a) found the pedagogical bridge he sought in the fact that the algebraic structure and order structure of positive scalar quantities — such as length, area, volume, and weight — are consistent with those of the real number system. Hence, work with the length, area, volume, and weight of real objects could enable children to appropriate the structural properties of the real number system, which Davydov considered the proper focus of school mathematics.

Another important factor in the construction of his curriculum was the role of dialectical logic which mandated that in order for a concept to be fully understood it must be traced from its origin through its developmental trajectory. In the case of mathematics especially this often implies an arduous process of historical and conceptual analysis followed by an equally lengthy psychological analysis in order to unpack, as it were, the full structure of a concept from the symbolic forms in which it has become embedded historically, and to recapitulate its developmental path in a manner that renders its appropriation accessible to children (cf. Davydov, 1990). Examples of the manner in which dialectical logic functions in the construction of Davydov's curriculum are provided below.

3 The Role of Dialectical Logic in the Development of Ordinal Number

An analysis of the manner in which the concept of ordinal number is developed is particularly illuminating of the role of dialectical logic in the structuring of the Davydov curriculum (Davydov *et al.*, 1999). Children are told that they are an ancient people who do not know how to count and one group is instructed to determine the length of a "mammoth" (a concrete object such as a rod of perhaps two or more yards in length is provided) and to convey this to a second group, but without numbers. The children have available various sticks of short length as well as small objects which could serve as tokens. This is a difficult task but eventually they decide to place a token for every time they lay off on the "mammoth" a length of the stick. In what to the best of our knowledge was the first implementation of Davydov's elementary curriculum in the U.S., the author challenged elementary teachers with this task prior to their using it with children, and in the process of conveying the tokens from the first group across an imaginary "mountain" to the second group, deliberately dropped a token without either group noticing. The second group, having been provided only with the tokens and small stick obtained from the first group, was charged with building a model of the mammoth. When this task was completed, they compared the length of the "mammoth" the second group had built with the original model. Of course, the second model was too short. Then the author "discovered" the dropped token, making the point that the number of tokens was so large that one had simply fallen as they were being conveyed and that a better method of representation was needed. In the ensuing discussion a suggestion to use tally marks in place of tokens was adopted. This model for the teaching of the problem followed the teaching recommendations of the Davydov program. It was then used by the elementary teachers in their own classrooms and with similar results. The children realized the disadvantages of tokens for cases of greater numerosity, and proposed using tally marks instead.

Soon after the classroom introduction of tally marks, a child was asked to leave the room and a length was placed on the board. A small stick functioned as a unit and tally marks replaced tokens as records. When the count was completed, the original length was removed and the child who left was called back into the room and given only the small stick and tally record which he successfully used to recreate the original length.

The teacher then asked for another volunteer to leave the room and repeated the previous problem placing a shorter length on the board and providing another very short length for the count. But this time the class decided to give the child the longer length together with the tally record when he reentered the room. The child assumed the length he had been given was the shorter unit length as occurred in the previous task, and of course, obtained the wrong result. The teacher pointed out that this ambiguity was intolerable and a better way of conveying the task requirements was needed. The class discussed this, proposed various expressions, and finally agreed to use the representation $U \rightarrow A$, where U represents the unit used and A represents the quantity. It was now possible not only to place tally marks above the arrow in $U \rightarrow A$ to express the number of units in a quantity A, but also to replace the U or A with a "?" to indicate whether the unit or the original quantity was to be determined. For example, three tally marks placed above the arrow would indicate that there were three units (U) in quantity A. If "?" appeared in place of the U and three tally marks appeared above the arrow, it would indicate that the quantity A was composed of three units, and it would be necessary to determine the unit. For example, in a simple case in which A was a rectangle consisting of three squares, one square would be seen to be the unit U. On the other hand, if U was designated by a square, and if "?" replaced quantity A and six tally marks appeared above the arrow, then the student was to build quantity A from six squares in any arrangement the student chose. Thus there were various ways to build a quantity given the designated unit. There was, however, only one possible determination of the number of units comprising a given quantity.

After working problems in which tally marks functioned well, a new problem was introduced that revealed their inadequacy. A long line segment was placed on the board together with a very small strip of paper. The children tracked the tally marks obtained from the count in their notebooks. But there were so many of these that the children had difficulty recording them accurately, so the teacher explained that ancient peoples sometimes adopted words for this purpose, and the children agreed that the words would have to occur in a particular order and be well-known to them in order to accomplish the required function. They decided that nursery rhymes would fill both purposes. Rather than tally marks, they could now write the word of the nursery rhyme that marked the end of a count above the arrow on $U \rightarrow A$. For example, in the nursery rhyme "Eenie, menie, minie, mo, catch a tiger by the toe...," if the number of units in the count was seven, rather than write seven tally marks above the arrow in $U \rightarrow A$, the children would write the word "tiger" above the arrow, since it marked the last word in the count.

Another challenging task, however, was soon to follow. In general nursery rhymes differ in Russian and in English traditions. Hence, we had to find English nursery rhymes to challenge the U.S. children with illustrations of the shortcomings of this new "nursery rhyme" method of designating a count. The children worked with various nursery rhymes. Then we presented them with the nursery rhyme "This little piggy went to market. This little piggy stayed home ..." Here the teacher observed that students used the word "piggy" to represent both lengths of three and nine. They concluded that this ambiguity was intolerable and therefore they could not use word sequences in which a word was repeated.

The children had now discovered three properties of ordinal numbers, viz. that they must be written in a particular order, that the order must be well-known (hence the choice to use nursery rhymes), and that no word could be repeated. The teacher next introduced a problem that resulted in too many units of count for the nursery rhyme to accommodate and the children concluded that the number of words must be sufficiently long to enable the recording of any count. They had now come to terms with the fourth property of ordinal numbers, viz. that they must be infinite, and were now ready to learn the names for numbers in common usage.

The introduction to ordinal number reveals the extent to which development of a concept is carefully traced from its origins through its developmental path by means of problems that render its appropriation by children possible. It is representative of the manner in which dialectical logic functions in the construction of the curriculum. The development of the ordinal numbers is also illustrative of the manner in which the Davydov curriculum develops topics through problems. The curriculum consists of a carefully designed and non-sectioned sequence of problems, and in this way follows the impetus for the development of mathematics in response to problems that arose historically. Its teaching method bears resemblance to that of constructivism, but emanates from a quite different theoretical position. In researching the manner of development chronicled in *Studies on the History of Behavior: Ape, Primitive, and Child*, Vygotsky and Luria (1993) noted that cognitive development occurred whenever a problem was encountered for which previous methods of solution were inadequate. Consequently, the method chronicled above, of confrontation with successive problems for which previously successful methods of solution break down, is characteristic of Davydov's curriculum, which has the cognitive development of students as its ultimate goal.

In a well-publicized study, Krutetskii (1976) had found that few students after having been taught the concept, could discern the structure of the square of a sum in expressions such as $(4x + y - a)^2$, 51^2, and $(C + D + E) \cdot (E + C + D)$. The vast majority needed many examples before the underlying structure became apparent to them. In analyzing Krutetskii's results, Davydov (1990) focused on the qualitatively different thinking of the few who were able to detect in these expressions the structure of the square of a sum even though the surface features of the expressions in question gave little indication of this structure. He realized that these students were thinking in a qualitatively different manner that enabled them to discern the structural essence beneath the deceptive forms, and he set out to discover whether it was possible to enable "ordinary" students to appropriate the theoretical mode of thinking displayed by the few who were successful with Krutetskii's tasks. Consequently, a major goal of his curriculum is developing in students the ability to think theoretically.

4 The Development of the Concept of Real Number from Quantity

The goal of theoretical learning and the role of dialectical logic are similarly evident in the manner in which Davydov's elementary curriculum begins, not with number, but rather with comparisons of quantities, a practice that antedates number in both cultural and individual histories. Children naturally compare lengths, volumes, etc., and in Davydov's curriculum (Davydov *et al.*, 1999), work problems requiring progressive comparisons in which finer distinctions must be made. They begin with the visual comparison of two quite disparate lengths and progress to the need to bring into alignment two objects of similar length in order to determine whether their lengths are equal. This is followed by requiring a comparison of two objects such as the length of a chalkboard and the height of a door, which cannot be brought into proximal alignment, but necessitate the introduction of an intermediary such as a rope in order to effect the required comparison. Eventually, children will have available only an intermediary of short length such as a stick, which then must be laid off on the objects in question, resulting in their *measures* which may be whole numbers or later fractions or irrationals.

Thus is number introduced after an initial semester of work with quantitative comparison, during which children also develop ways to produce an equality given two objects of unequal length or volume by either adding to the lesser or subtracting from the greater the difference between them. Since children work extensively with quantities for which they as yet have no numerical measures, they label the quantities with letters, and as their actions with quantities become more refined and complex over an entire semester, they are expressed algebraically. Thus algebraic development precedes numerical applications, providing another example of the ascent from the abstract to the concrete (Davydov, 1975b; Minskaya, 1975; Schmittau, 2005).

If A and B are the greater and lesser of two volumes, respectively, children may add a volume C to B or pour out a volume C from A, thereby transforming the inequality $A > B$ into equalities $A = B + C$ and $B = A - C$, where $C = A - B$. They employ a representational schematic "$\wedge$", placing A (the whole) at the apex of the "$\wedge$" and B and C (the parts), at the bottom of each of the line segments as follows:

$$\begin{array}{c} A \\ / \backslash \\ B \quad C \end{array}$$

The schematic functions as a *psychological tool*, focusing the children's analysis on the part-whole structure of the quantities in question. In addition, it provides a record of the relationship of the quantities throughout children's transformative actions on them.

Unlike material tools that effect action on the outer world, psychological tools function to direct attention inward to the control of behavior, which in this case is the transformation of the quantities in question through their own activity, while maintaining their focus on the theoretical part-whole structure of the quantities. Although the ultimate goal is the development of the real number system, dialectical logic is in evidence as the origins of the concept of number are traced back to the comparisons of the properties of length, area, volume, and weight of physical objects, the progressive refinement of which will result in the definition of real number as the outcome of their measurement.

In this brief sketch of the beginnings of the children's study, the complementary and synergistic roles of theoretical learning, dialectical logic, and psychological tools are evidenced. Theoretical learning requires discernment of the essence of a concept, dialectical logic mandates that its development be traced from its incipience through to the highest level of its present attainment, and the psychological tool serves to direct attention of the learner to the underlying theoretical structure of the concept under consideration.

5 Unresolved Issues and the Manner of Their Resolution by Davydov

The various movements in mathematics education in the U.S. throughout the second half of the 20th century experienced challenges mentioned in Section 1 of this monograph. Among these were those identified by psychological research that revealed differences in mathematics usage by practitioners in the workplace and market place vs. students whose main usage of mathematics was confined to classroom exercises (Carraher, Carraher, and Schliemann, 1985; Saxe, 1997; Scribner, 1997). The former evinced flexibility in applying mathematics to the solving of problems arising in market and work settings, while the latter employed rigidly singular algorithmic solution methods to such problems (Scribner, 1997). Questions inevitably arose concerning the potentially remediating effect of introducing into the classroom problems couched in real world settings, and whether all learning was, in fact, context bound and attempts to teach the abstractions of mathematics were doomed to be narrow and ineffectual. These issues occupied mathematics educators during the 1980s, but were far from resolved when they dissolved into the push for curricular reform at the end of the decade.

Although the U.S. reform movement was concerned with developing flexibility in mathematical thinking and problem solving, the aspect of iterative or repetitive practice that was essential to such development was considerably downplayed, as calculators were introduced into the elementary school and replaced children's former mental computational practice. Their resulting inability to perform even simple computations is reflected in data recently gathered from high school students who used calculators for computations such as $9 - 5$, $6 - 2$, 3×3, $6 + 4$, and $21 - 10$. It is further evidenced, for example, by the elimination at the secondary level of factoring by inspection and completion of the square. Since students have not mastered the requisite knowledge of simple sums and products, they find it impossible to discern the requisite factors and sums required for factoring trinomials such as $x^2 + 2x + 1$ or $x^2 + 7x + 12$. Instead they employ the necessarily postulated quadratic formula for these simple factoring tasks. This is a predictable consequence of the formal logical division of concepts, problem solving, and computation into separate domains and the privileging of concept development and problem solving over computation throughout the last decade. Without foundational procedural knowledge the ability to solve problems flexibly is compromised.

Russian psychologists also promoted flexibility of mathematical thinking, and in what to our knowledge is the first implementation of Davydov's curriculum in a U.S. school setting, children consistently were observed solving problems such as $13 + 17$ in a variety of ways such as those depicted in Figure 1 (cf. Schmittau, 2004 for an extensive discussion).

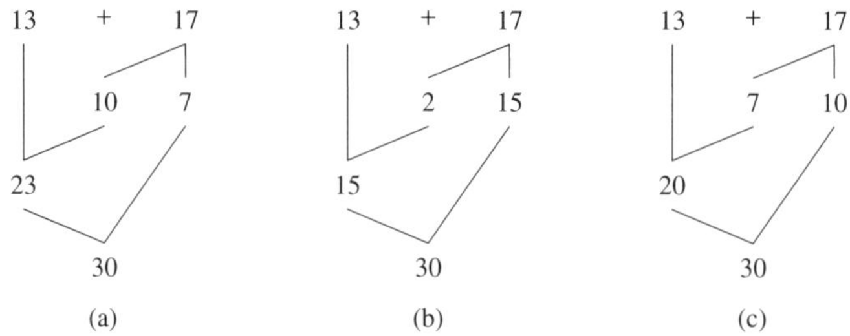

Fig. 1. Example of children's solutions to an addition problem.

In the first example in Figure 1(a), the child thought of 17 as composed of the parts 10 and 7, and then added 13 and 10, obtaining 23 for the sum. She then added the 23 plus 7 to obtain 30. In the second example (Figure 1(b)), the child first decomposed 17 into parts 2 and 15, and then added 13 and 2 to obtain 15. Finally he added 15 plus 15 to obtain the sum of 30. In the third example (Figure 1(c)), the child decomposed 17 into parts 7 and 10, added 13 to 7, and then added the resulting sum of 20 to 10 to obtain 30. The solutions presented in Figure 1 indicate not only flexibility of thinking but also incorporate the iterative process so essential to computational mastery (Lave, 1988). The U.S. reform movement has done much to encourage similar flexibility, but mental computational practice has been curtailed by the introduction of calculators in the elementary grades.

In the Davydov curriculum mental computational or iterative practice is ongoing, and as multiplication is introduced in the second grade, children continue to develop facility in computation simultaneously with their ability to solve problems. An example is given in their work to build the multiplication tables (Davydov *et al.*, 2000). After developing the tables for multiplication by 2 and 3, multiplication by 4 is challenging, and a problem such as 4×7 is anything but a routine exercise. A child might solve it by employing the distributive property (developed earlier in work with quantities) to figure out the product:

$$4 \times \underset{2 \;\; 5}{\underset{/\backslash}{7}} = (4 \times 2) + (4 \times 5) = 8 + 20 = 28.$$

Or 7 might be split into 4 and 3, or 4 into 2 and 2.

Since the commutative property has been developed and the child has found the products of 2 and of 4×5 prior to confronting 4×7, those results may be used in finding a solution. And such a problem, far from being routine, requires the child to think about possible ways to use what is known to generate the product. Work continues over time with progressively more difficult problems such as the following:

$$9 \times 16 = 9\,(10 + 6) = (9 \times 10) + (9 \times 6) = 90 + 54 = 144$$

/ \

10 6

$$8 \times 935 = 8\,(900) + 8\,(30) + 8\,(5) = 7200 + 240 + 40 = 7200 + 280 = 7480.$$

/ \ \

900 30 5

Vertical representations of the solution process can then be introduced and expanded to encompass the multiplication algorithm for any two numbers no matter how large (cf. Schmittau, 2004 for a more extensive discussion).

The distributive property of division is also studied and applied in a similar manner in Davydov's curriculum, again with extension to the division algorithm. The key concepts are the part-whole structure of any quantity or number, the distributive property of positive scalar quantities and their numerical designations, and the concept of positional system. These elements are present and rendered explicit throughout the entire development of multiplication and division, from the beginnings of work with actions of multiplying and dividing through to their consummate development in the respective algorithms. The "∧" schematic functions as a psychological tool to direct attention to the theoretical structural essence, the fact that numbers may be broken into parts, thereby allowing for computation by invoking only a relatively small number of memorized products or sums. And children discover as the work progresses that it is often easier to accomplish by breaking large numbers into the parts indicated by their position within a positional system. Consequently, the many ways in which children may solve problems initially, such as the addition and multiplication problems presented above, eventually resolve into the most efficient computation methods of partitioning the numbers in question along the lines of the number of units, of bases (such as tens), of squares of bases (such as hundreds), etc. Positional systems of various bases are extensively studied during the second semester of the first grade curriculum.

Children are also required to work at theoretical levels with positional systems in the first grade, solving such problems as inserting the correct order symbol ($<$, $=$, or $>$) into an expression such as $\Delta\partial 0$? $\Delta 0\partial$. Here symbols other than Hindu–Arabic numerals are used. Children must notice that regardless of the numerals designated by Δ and ∂, the placement of the zero in the respective numerical expressions implies that the first number is greater than the second, since $\partial > 0$ in the second position (which in base 10 would be the tenths position). Hence, they must understand the concept of positional system at an abstract level. Children also are asked to order on a number line numbers expressed in various bases (cf. Schmittau, 2003a, in press, for more extensive discussion).

As a result of the development outlined above, equations and inequalities are not perceived by students as separate topics to be studied successively, nor are addition and subtraction or multiplication and division separate operations. Rather, equations and inequalities arise through the transforming *actions* of adding or subtracting and these actions occur simultaneously within the context of changing quantities to render them equal or unequal. Dialectical logic goes to the origin of a concept and identifies its essence, and it is this *essence* that forms the connections that exist and persist throughout the developmental trajectory of the concept, the highest elementary levels of which are often the powerful culturally constructed algorithms capable of operating on any numbers no matter how large. In this way the apparent dichotomy between the conceptual and procedural dissolves as the procedural is seen to be fully conceptual (Schmittau, 2004).

Moreover, the many problems that must be solved in order to fully develop a concept such as multiplication or division perform the iterative function that is characteristic of "workplace" mathematics (Lave, 1980), and thereby develop the flexibility of thinking common to mathematical usage acquired in informal settings. Problem solving ability is developed at the same time, since even a seemingly simple computational problem such as 4×7 presents a challenge to the child who only knows multiplication by 2 and 3. There is then, no trade-off between computation and problem solving, since computation *is* problem solving at every step of the way and is, in addition, infused with conceptual content. These abilities develop together as it might be imagined that they occurred historically in response to societal and economic situations of increasing complexity and numerosity.

In our experience implementing Davydov's curriculum in the U.S., children consistently provided conceptual arguments for computational solutions, and by the third grade level accurately solved problems such as $6080 \cdot (145666/173 - 88508/116)$ without the need for calculators (Schmittau, 2004, p. 33).

6 The Development of Mathematics as a Conceptual System

Since the Davydov curriculum was in progress decades before the Carraher, Carraher, and Schliemann (1985), Saxe (1997), and Scribner (1997) studies, the question naturally arises as to what accounted for building into it the iterative development so characteristic of mathematics learned in the workplace. The answer is contained in Vygotsky's (1986) commentary on conceptual systems in which he speaks of the many ways in which a number such as "1" can be expressed as the difference between any integer and its predecessor or the quotient of any number divided by itself. Ironically, by consigning the simplest of computations to a calculator early in the elementary school years, U.S. children are deprived of the attainment of the flexibility the curriculum ostensibly seeks to promote. This flexibility must be developed by the challenge of solving numerous computational problems requiring thought about the many ways any number can be expressed and the numerical relationships into which it enters within the whole number, integer, rational, and finally real number systems. It is not sufficient that a calculator can produce a computational answer, for it is not the *answer* that is of the greatest importance to elementary school children's mathematical development. Rather it is their ability to apprehend mathematics as a *conceptual system*. Children can no more do this without mental computational practice than they can attain effective verbal expression without a sufficient vocabulary and knowledge of the interrelationships among words — their equivalence or near equivalence, shades of meaning, etc., all of which are attained through extensive practice with the oral and written usage of the language.

7 The Role of Dialectical Logic and Psychological Tools in Developing Multiplication and Division

Another thread of the present discussion deserves further commentary. Multiplication is taught in the U.S. as repeated addition, thereby reinforcing the generative metonymic role of counting number in the real number system (Schmittau, 2003b). By way of contrast, dialectical logic impels examination of the actual cultural conditions under which not only the concept of number but the action of multiplication may arise. Davydov (1992) cites Lebesgue (1960), and following his tradition, approaches multiplication as a change in the system of units. This change occurs in any circumstance in which it is required to take a count or measure of many units, so that their very numerosity presents a daunting task and one that is fraught with the possibility of error.

In their classroom experience children may be required to determine how many small cups can be filled from a large pitcher of juice. A number of larger glasses are placed on the table along with the pitcher and small cups, but no mention is made of them. As children continue filling the small cups from the pitcher, the task becomes tedious and the suggestion will be made that they fill the larger glasses and then determine how many of the small cups a large glass will fill. Hence, the unit has been changed from the small cup to the larger glass. If four small cups fill a large glass and six large glasses fill the pitcher, then an indirect count of the number of cups the pitcher will fill can be obtained by noting that the number will be four taken six times. This is the origin of the *concept* of multiplication, which later will be expressed as 4×6 (cf. Davydov, 1992, for an extensive discussion). Multiplication then, is not reduced to a form of addition as is the common practice in U.S. textbooks, but in its conceptual *origin* is a different mathematical action altogether.

This change in the system of units also requires a reconsideration of the definition of unit, for it challenges as inadequate the designation of unit as a single discrete object. The concept of unit is thus seen to be antecedent to the concept of multiplication and to the concept of measurement as well. In the second grade curriculum of Davydov, for example, children work many problems in which they must either build or measure with units which are not simply single discrete entities, but instead may be composed of one or more shapes of various types (Davydov *et al.*, 2000).

The psychological tool employed is $U \to A$, where U represents the unit and A the quantity to be built or measured. The following sign represents that a quantity containing six units is to be built from U.

$$U \xrightarrow{6} ?.$$

The sign below represents a quantity A to be measured with the unit U.

$$U \xrightarrow{?} A.$$

In representing multiplication, where a change in the system of units occurs, this representation or psychological tool must be adapted to reflect the change in units. Consequently, in the Davydov curriculum a schematic is drawn in which the elemental unit (U) is changed to a composite unit. The composite unit is then multiplied by the number of composite units to form the product. In the example above, the small cup is the elemental unit, while the glass is the composite unit, since it is composed of four elemental units. Consequently, the schematic drawn must reflect the change from a single elemental unit (cup) to a composite unit of four cups. This composite unit must then be shown to be multiplied by six to obtain the product of $4 \times 6 = 24$ cups in the pitcher, a result that is obtained without measuring out each individual cup.

Such a schematic can also be used to represent both measurement and partitive division, and in so doing reflects the fact that, just as in adding and subtracting, multiplying and dividing are not separate and formally connected only as inverse operations, but are, in fact, connected through their origins in activity. The following problems typical for Davydov's curriculum illustrate both this and the role of the schematic in representing the relationship between multiplication and the two types of division.

Consider the following problem: Five children are to each receive a small bag containing 3 pieces of candy. How many pieces of candy are required? Clearly this is a multiplication problem whose solution is $5 \times 3 = 15$. Modifying the problem to one of measurement division would require asking how many bags of candy comprised of 3 pieces in each bag can be made from 15 pieces of candy. Further modification to produce a problem requiring partitive division would involve reframing the problem as follows: If 15 pieces of candy are to be distributed evenly among 5 children, how many pieces will each child receive? The schematics for each of the three related problems appear in Figure 2.

In the multiplication instance in Figure 2, the composite unit C (3 pieces of candy) is formed from the elemental unit U (a single piece of candy). Five such composite units give the product $5 \times 3 = 15$. In the measurement division instance in Figure 2, the composite unit $C = 3$ must be divided into 15 to yield the quotient of 5 bags of candy composed of 3 pieces each. In the instance of partitive division shown in Figure 2, 15 must be divided by 5 to yield the quotient of 3 pieces in each bag, which is the composite unit C formed from

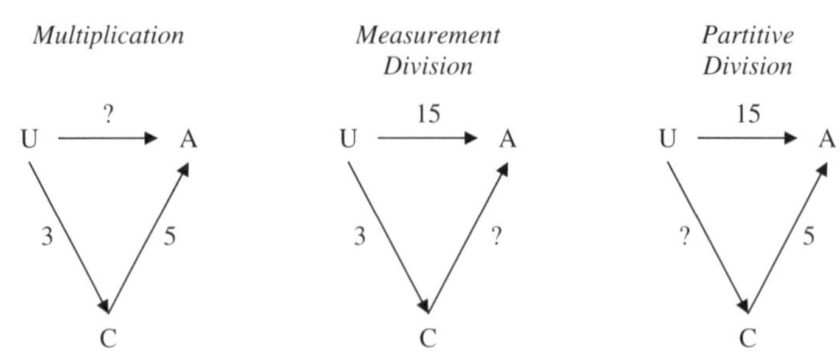

Fig. 2. Schematic for multiplication and division.

the elemental unit U. The interrelationships between the three types of actions are readily discerned. Consequently, multiplication and division are connected at their origins in activity rather than as separate or formally inverse operations. Thus these apparently disparate operations are united in their dialectical complementarity and this is reflected in the psychological tool or schematic that serves a representational function. The diagram also facilitates internalization of the *essence* of the action and its interrelationship with related actions.

8 Problematizing the Understanding of Unit

It is immediately obvious why it is necessary to problematize the understanding of unit beyond that of a single discrete object. The extensive practice with building and measuring with composite units prior to the introduction of multiplication in Davydov's curriculum serves the function of familiarizing the student with the concept of unit as flexible rather than rigidly bound to a single discrete entity (Davydov *et al.*, 2000). Not only is this essential for multiplication in which a single or elemental unit is changed to a composite of several such units as in the problem above, but it is also necessary for the understanding of measurement. Children using Davydov's curriculum in the U.S. were thrilled to discover that they could find out how many milligrams were in a kilogram by simply replicating the schematic for multiplication, a task that would have been considerably more unwieldy to perform with actual measurements.

The consequences of failure to master measurement systems, which arise in many areas of practical application, can be anything but trivial. Vagliardo (2008), for example, found inadequate conceptualization of measurement to be widespread among nursing students, and measurement conversion errors in medical practice have been well documented in the literature. Indeed the term "death by decimal" (Przybycien, 2005, p. 32) has been applied to the failure to detect the consequences of a 10-fold dosage error resulting from a misplaced decimal in a conversion involving milligrams of a prescribed drug. Such failure has the potential to render the dosage either totally ineffectual or lethal, depending on the direction of the error.

9 Avoidance of Generative Metonymy

It is also important to note that the approach taken not only to number but also to multiplication in the Davydov curriculum avoids the conceptualization of number as a generative metonymy, and its metonymic reinforcement by such fundamental operations as multiplication (Schmittau, 2003b). When number is generated out of counting as is the case in U.S. schools, the counting numbers serve the function of a generative metonymic. In generative metonymies an entire category is produced from a subset of members of the class, and this is the case in the formal derivation of the rational and real numbers from the counting numbers, wherein the rational numbers are defined as quotients and the real numbers as sequences of counting number digits (Lakoff, 1987). In addition, students tend to conceptualize the real numbers and their algebraic forms together with the operations performed on them in terms of counting numbers, even when these are clearly inadequate to represent the full range of real numbers. The defining of multiplication as repeated addition extends this role, since a factor can only be added to itself a counting number of times. This is illustrated in a comparative study in which U.S. high school and university students, in contrast to their Russian elementary school counterparts, conceptualized the monomial product "$a \cdot b$" as the product of small whole numbers. Russian children who had completed only the three years of Davydov's elementary curriculum understood the factors "a" and "b" in their generalized nature as representations of *any* numbers (Schmittau, 1994, 2003b).

10 The Role of Dialectical Logic in the Development of Psychological Tools

The above discussion of the ordinal numbers presents an example of the manner in which dialectical logic functions even in the development of psychological tools, as not only the need but also the requirements for the representational schematic $U \rightarrow A$ are explored in the classroom setting within Davydov's curriculum. This representation is extended as noted above, when multiplication is introduced, and is not only adaptable to the representation of both measurement and partitive division, but also capable of reflecting their dialectical complementarity through their origins in activity.

Such development is typical of other psychological tools as well. The number line, for example, arises in Davydov's curriculum from a consideration of simple medicinal dosage calculations using a graduated cylinder which is then tipped ninety degrees to create a horizontal gradient. The elements essential to the creation of a number line are also explored thoroughly, including direction, starting point, and choice of a unit. If any one of these is unspecified, it is impossible to determine where specific numbers will be located even if the other two elements are provided. When the number line is presented as a ready made representation with these elements simply in place *a priori*, as is typically the case in the U.S., the arbitrary nature of these determinations remains undetected, since they are never explored. When children are presented with a number line such as in Figure 3, and asked to mark "$a + 1$" on the number line, they must notice that no unit has been provided and therefore it is impossible to complete this task. Similar problems appear in the Davydov curriculum with respect to other elements essential to the number line that may be missing, such as beginning and direction. These include problems that function as "traps," having no solution because of missing or contradictory information. "Traps" are distributed liberally throughout the curriculum, so that it can never be assumed that any problem with which a student may be confronted is in fact capable of a solution.

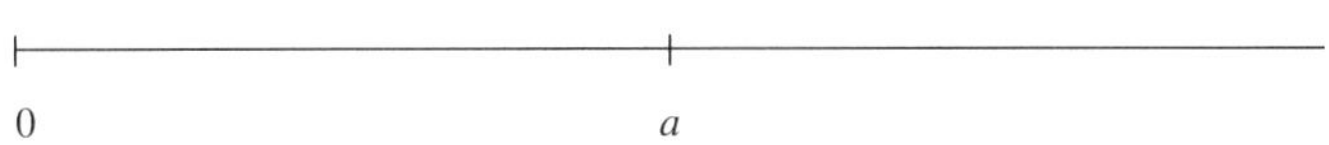

Fig. 3. Number line.

Even in the case of the common "table," a psychological tool of the first order, there is a full development — far too lengthy to permit its chronicling here — beginning with the need to arrange minimal information presented in a narrative format with no mathematical question asked. This process continues through greater levels of complexity in the Davydov third grade curriculum, until it culminates in the development of a tabular representation capable of use in problems requiring proportional reasoning and at the same time, extendable across a wide range of problem types and situations (Davydov *et al.*, 2001). Children even create tables to organize literal designations for geometric dimensions which they use to facilitate the solving of problems for which no numerical data are provided (Schmittau, 2004, in press).

11 The Role of Dialectical Logic in Developing Effective Classroom Participation

The extent to which dialectical logic and conceptual analysis are applied in the Davydov program can be seen also in the manner in which even children's ability to respond appropriately within the context of a classroom becomes the subject of explicit learning. Social competence presupposes the ability to detect and interpret often subtle verbal and non-verbal cues signaling different participation structures, and children who miss such signals within a classroom context fail to adapt their behavior accordingly and often act inappropriately in the new situation (Erikson and Schultz, 1997). Russian researchers moved proactively to address the failure of young children to respond appropriately to transitional cues and to familiarize them with the requirements for effectively working within various class structures.

A 10-day course, *Introduction to School Life* (Zuckerman and Palivanova, 1992), consisting of 30 class hours, was designed to engage children in role playing and dialogues using stuffed animals as story characters. These require children to analyze the mode of appropriate participation in whole class, small group, dyadic, and individual work. They also render explicit the kinds of cues that signal changes in participation structure and even the kinds of question forms that may indicate whether a group or an individual response is being called for. By teaching the course just prior to or at the beginning of the first grade, Russian educators "level the playing field" for young children, who initially differ substantially in their ability to detect and respond appropriately to the various cues implicit in the verbal and non-verbal behavior that characterizes the school contexts they are about to enter.

It is even less likely that young children will know how to conduct themselves in productive ways in the variety of new group learning situations they will experience; hence, both appropriate and inappropriate models are presented for analysis. Consequently, children role play debates between the animal characters who may be egotistical, unsure of themselves, or too trusting. They find that any spurious claims to superiority or appeals to authority are quickly disregarded. Only arguments supported by reasoning offered in support of their position are acceptable (Zuckerman, 1994). Research confirmed that children participating in the program demonstrated greater ability to coordinate their actions, engage in cooperative planning, obtain missing information, settle conflicts amicably, and accurately appraise their own and their peers' performance (Zuckerman, 1994). In my research in Russian schools using Davydov's curriculum I consistently found that even first graders disagreed with their classmates without becoming disagreeable and demonstrated the ability to mount an argument and pursue it logically. In general, they behaved like responsible adults in problem solving situations.

12 Reflections/Conclusion

The development of various major concepts in the Russian elementary curriculum of Davydov reflect, as indicated above, the role of dialectical logic, the ascent from the abstract to the concrete, and the role of psychological tools in the promotion of theoretical learning that is essential not only to a deep understanding of mathematics but to the cognitive development of students as well. These factors assume powerful functions within the curricular content. And the instructional methodology, although considered by Vygotsky to be less important than and incapable of substituting for the theoretical content, also functions to promote cognitive development.

In addition, the role of dialectical logic is found not only in the development of the content of the elementary mathematics curriculum, but also in the manner in which psychological tools are introduced and mastered, and even in the explicit attention to the student's role in appropriate interactions within the classroom culture. It is dialectical logic that functions to overcome many of the apparent dichotomies listed in the opening paragraphs of this monograph and that have yet to be fully overcome in U.S. classrooms. And it is the employment of psychological tools that overcomes the objections the author hears so often from U.S. teachers, namely, that students can "work the problems with manipulatives, but not with numbers." The psychological tool bridges the gap between action on objects and action on numerical or algebraic forms, and its absence results in a perceived lack of congruence between such actions, rather than discernment of the latter as merely a symbolic trace of the former.

In the interest of promoting the connections so essential to the understanding of mathematics as a conceptual system, current approaches to resolving the apparent multiple dichotomies mentioned above tend (again within the framework of formal logic) to address their resolution one by one. This involves of necessity a lengthy process, confronting mathematics education with an array of issues with which to occupy itself for some time to come. At the same time it allows mathematics education to avoid confronting the more far reaching perspectives that give rise to these apparently disparate entities. It is here that Russian elementary education is of particular relevance. The approach taken by Davydov is more efficient, but it requires "laying the axe to the root of the tree," as it were, and thinking deeply about the origins of these seemingly disparate concepts and dichotomies and the manner in which they have been culturally created. This, in turn, necessitates a consideration of their alternative development through dialectical logic, a perspective within which they simply do not arise.

References

Bourbaki, N. (1963). *Essays on the History of Mathematics [Oeherki po istoril matematiki]*. Moscow: Inostrannaya literatura.

Carraher, T. N., Carraher, D. W., and Schliemann, A. D. (1985). Mathematics in the streets and in schools. *British Journal of Developmental Psychology*, *3*, 21–29.

Davydov, V. V. (1975a). Logical and psychological problems of elementary mathematics as an academic subject. In L. P. Steffe (Ed.), *Soviet Studies in the Psychology of Learning and Teaching Mathematics*, Vol. VII (pp. 55–107). Chicago: University of Chicago Press.

Davydov, V. V. (1975b). The psychological characteristics of the "prenumerical" period of mathematics instruction. In L. P. Steffe (Ed.), *Soviet Studies in the Psychology of Learning and Teaching Mathematics*, Vol. VII, pp. 109–206. Chicago: University of Chicago Press.

Davydov, V. V. (1990). *Types of Generalization in Instruction: Logical and Psychological Problems in the Structuring of School Curricula.* Reston, VA: National Council of Teachers of Mathematics (Original published in 1972).

Davydov, V. V. (1992). The psychological analysis of multiplication procedures. *Focus on Learning Problems in Mathematics, 14*(1), 3–67.

Davydov, V. V., Gorbov, S. F., Mikulina, G. G., and Saveleva, O. V. (1999). *Mathematics: Class 1.* Edited by J. Schmittau. Binghamton, NY: State University of New York.

Davydov, V. V., Gorbov, S. F., Mikulina, G. G., and Saveleva, O. V. (2000). *Mathematics: Class 2.* Edited by J. Schmittau. Binghamton, NY: State University of New York.

Davydov, V. V., Gorbov, S. F., Mikulina, G. G., Savyelyeva, O. V., and Tabachnikova, N. L. (2001). *Mathematics: 3rd Grade.* Edited by J. Schmittau. Binghamton, NY: State University of New York.

Elkonin, D. B. (1975). Primary schoolchildren's intellectual capabilities and the content of instruction. In L. P. Steffe (Ed.), *Soviet Studies in the Psychology of Learning and Teaching Mathematics*, Vol. VII (pp. 13–54). Chicago, IL: University of Chicago Press.

Erikson, F., and Schultz, J. (1997). When is a context? Some issues and methods in the analysis of social context. In M. Cole, Y. Engeström, and O. Vasquez (Eds.), *Mind, Culture, and Activity* (pp. 22–31). Cambridge, UK: Cambridge University Press.

Hegel, G. (1956). *Sochineniya* [*Works*], Vol. 3. Moscow: Gozpolitizdat.

Kozulin, A. (1990). *Vygotsky's Psychology: A Biography of Ideas.* Cambridge, MA: Harvard University Press.

Kozulin, A. (1998). *Psychological Tools: A Sociocultural Approach to Education.* Cambridge, MA: Harvard University Press.

Krutetskii, V. A. (1976). *The Psychology of Mathematical Abilities in School Children.* Chicago: University of Chicago Press (Original published in 1968 in Russian).

Lakoff, G. (1987). *Women, Fire, and Dangerous Things: What Categories Reveal About the Mind.* Chicago: University of Chicago Press.

Lave, J. (1988). *Cognition in Practice.* New York: Cambridge University Press.

Lektorsky, V. A. (1984). *Subject, Object, Cognition.* Moscow: Progress Publishers.

Minskaya, G. D. (1975). Developing the concept of number by means of the relationship of quantities. In L. P. Steffe (Ed.), *Soviet Studies in the Psychology of Learning and Teaching Mathematics*, Vol. 7. Chicago: The University of Chicago Press.

Przybycien, B. (2005). *Safe Meds: An Interactive Guide to Safe Medication Practice.* St. Louis: Elsevier Mosby.

Saxe, G. B. (1997). Selling candy: A study of cognition in context. In M. Cole, Y. Engeström, and O. Vasquez (Eds.), *Mind, Culture, and Activity: Seminal Papers from the Laboratory of Comparative Human Cognition* (pp. 330–337). Cambridge, UK: Cambridge University Press (Original published in 1989).

Schmittau, J. (1993). Vygotskian scientific concepts: Implications for mathematics education. *Focus on Learning Problems in Mathematics, 15*(2–3), 29–39.

Schmittau, J. (1994). Scientific concepts and pedagogical mediation: A comparative analysis of category structure in Russian and U.S. students. Paper presented at the Annual meeting of the American Educational Research Association, New Orleans, LA, April, 1994.

Schmittau, J. (2003a). Beyond constructivism and back to basics: A cultural historical alternative to the teaching of the base ten positional system. In B. Rainforth and J. Kugelmass (Eds.), *Curriculum and Instruction for All Learners: Blending Systematic and Constructivist Approaches in Inclusive Elementary Schools* (pp. 113–132). Baltimore, MD: Brookes Publishing Co.

Schmittau, J. (2003b). Cultural historical theory and mathematics education. In A. Kozulin, B. Gindis, S. Miller, and V. Ageyev (Eds.), *Vygotsky's Educational Theory in Cultural Context* (pp. 225–245). Cambridge, UK: Cambridge University Press.

Schmittau, J. (2004). Vygotskian theory and mathematics education: Resolving the conceptual–procedural dichotomy. *European Journal of Psychology of Education, IX*(1), 19–43.

Schmittau, J. (2005). The development of algebraic thinking: A Vygotskian perspective. *Zentralblatt Für Didaktik Der Mathematik, 37*(1), 16–22.

Schmittau, J. (in press). The role of theoretical analysis in the development of algebraic thinking.

Scribner, S. (1997). Mind in action: A functional approach to thinking. In M. Cole, Y. Engeström, and O. Vasquez (Eds.), *Mind, Culture, and Activity: Seminal Papers from the Laboratory of Comparative Human Cognition* (pp. 354–368). Cambridge, UK: Cambridge University Press (Original published in 1992).

Vagliardo, J. J. (2008). *Mathematics and nursing: Students' conceptual understanding of mathematics for nursing.* Doctoral Dissertation. State University of New York at Binghamton.

Vygotsky, L. S. (1986). *Thought and Language.* Cambridge, MA: MIT Press (Original published in 1934).

Vygotsky, L. S., and Luria, A. R. (1993). *Studies on the History of Behavior: Ape, Primitive, and Child.* Hillsdale, NJ: Erlbaum.

Zuckerman, G. A. (1994). A pilot study of a ten-day course in cooperative learning for beginning Russian first graders, *The Elementary School Journal, 94*(4), 405–420.

Zuckerman, G. A., and Palivanova, K. N. (1992). *Introduction to School Life.* Moscow: Novaya Shkola.

8

The Preparation of Mathematics Teachers in Russia: Past and Present

Natalya Stefanova

Herzen State Pedagogical University of Russia, Russia

1 Introduction

An orderly system of preparing teachers of mathematics has evolved in Russia. Its principal motto may be formulated as follows: "Professionalism in a mathematics teacher = fundamental education + professional competence." Mainly due to the consequences of Russia's endorsement of the Bologna Declaration (Bologna, 1999), significant changes are taking place today in the Russian system of mathematics teacher preparation, and in the Russian system of professional education as a whole as well. In addition, the distinctive features of mathematics teacher preparation in Russia cannot be understood without a description of the processes taking place in general secondary mathematics education, i.e., in the future sphere of the mathematics teacher's professional activity. It is this sphere that dictates the requirements for teacher preparation, and therefore, it is this sphere that determines the content and structure of such preparation. Therefore, we will first turn to a brief description of the system of mathematics education and the contemporary tendencies that are shaping its development.

2 The System and Content of General Secondary Mathematics Education in Russia Today

Mathematics education is a mandatory part of the official system of continuous education in Russia. The objectives and requirements that students in mathematics are expected to meet are determined by official guidelines, which include official education standards (e.g., Government Education Standards. Mathematics, 2004), subject curricula (e.g., The Mathematics Curriculum, 2002), and other regulatory documents (e.g., The Required Content Minimum, 1999).

As an element in the system of continuous education, mathematics education is offered at all stages of education: preschool education (in preschool institutions), general education (in general education schools), and professional education (as part of secondary education, in "colleges"[1] and lyceums; and as part of higher education, in institutes, academies, and universities). Thus, one may distinguish between preschool mathematics preparation, general mathematics education, and secondary and higher professional mathematics education (Diagram 1).[2] It should be noted that the only stage in the educational system that is universally mandatory is general education. Preschool mathematics preparation is acquired only by those children who attend preschool institutions, principally kindergartens. Continuing education after the general education school also is not mandatory.

Secondary professional education includes a complete secondary general education (including mathematics education), as well as specialized education in a subject related to a selected profession or group of professions. For this reason, general and secondary professional mathematics educations overlap in the diagram above.

A particularly important role in the system of mathematics education is played by general mathematics education. It is at this stage

[1] Note that this word is currently used in Russia to denote institutions that offer vocational secondary education.

[2] In addition to the educational systems represented in the diagram, there also is a supplementary system of mathematics education in Russia, which is oriented toward working with mathematically gifted children.

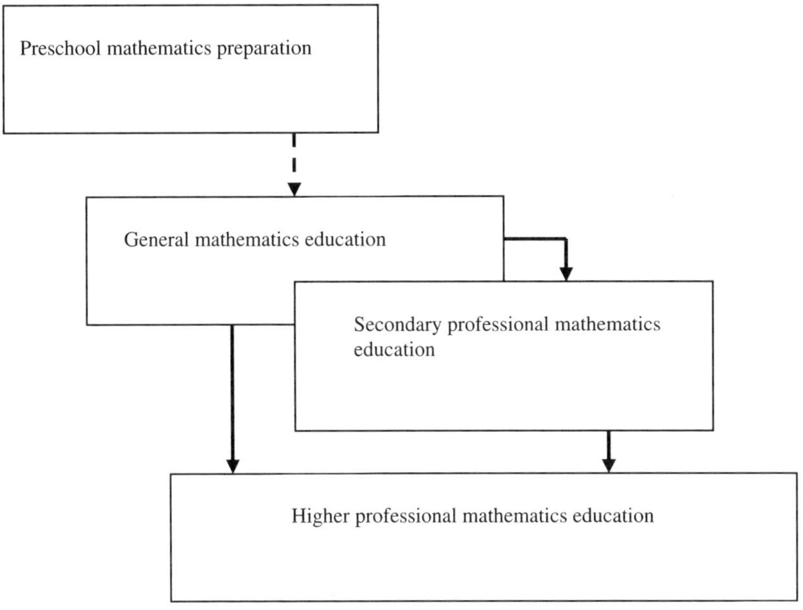

Diagram 1.

that the foundations of mathematical knowledge are laid and a basis is formed on which future preparation programs can be built.

As can be seen in Diagram 2, the model of general mathematics education also can be represented as a three-stage structure, comprising elementary, basic, and complete (secondary) education.

This structure corresponds to the structure of the system of general education: elementary education (elementary school), basic education (basic school), and complete secondary education (senior or high school). Elementary schools are attended by children of ages 7–10, basic schools by children of ages 11–15, and senior schools by children of ages 16–17.

Today, a system of specialized classes is employed at the stage of the senior (high) school (Profiled, 2004): students may select among different educational programs in accordance with their inclinations and interests. In each of the available programs, mathematics may be studied either at a specialized or a basic level. The contents of the basic and specialized courses in mathematics differ in terms of the volume of

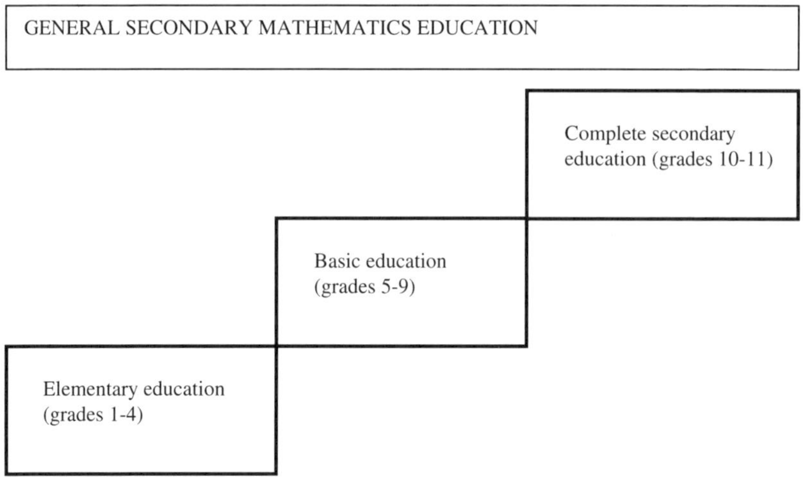

Diagram 2.

the information covered, the degree to which mathematical propositions are substantiated, and the difficulty of the problems given to the students. The number of hours allocated for the basic and specialized courses in mathematics is typically 4 and 6 hours per week, respectively. In addition, over two years of schooling in the upper grades, students may choose one or several elective courses in mathematics, aimed at supplementing and reinforcing the specialized course.

In addition to the basic and specialized levels of mathematics education, there is also a level called "advanced study." Students who choose to follow this track have at least eight mandatory hours of mathematics per week.

The contents of mathematics education in the general school is distributed among the following subjects: mathematics (grades 5 and 6); algebra (grades 7–9), geometry (grades 7–9); and in grades 10 and 11 at specialized level: algebra and elementary calculus, and geometry, at basic level: mathematics.

The contents of these classes can be described as follows.

Mathematics (grades 5 and 6) — sets of natural numbers, fractions (positive), and rational numbers; expressions using variables and their simplest transformations; elementary linear equations; word problems; certain geometric figures and their properties; elementary probability theory.

Algebra (grades 7–9) — the set of real numbers; algebraic expressions (including fractional algebraic expressions), expressions with square roots and their equivalent transformations; linear quadratic functions; the proportional relation; certain cases of the power function; solving algebraic equations and inequalities, and solving systems of algebraic equations and inequalities using equivalent transformations; arithmetic and geometric progressions; elementary probability theory and statistics.

Geometry (grades 7–9) — Euclidean plane geometry: axioms, triangles, and their congruence; quadrilaterals; polygons; the areas of polygons; circles and inscribed (circumscribed) polygons; geometric transformations in the plane; vectors and coordinates in the plane; elementary trigonometry.

Algebra and elementary calculus (grades 10 and 11) — transcendental functions on the set of real numbers (power, exponential, logarithmic, trigonometric, inverse trigonometric functions); transcendental equations and inequalities; derivative and anti-derivative; elementary probability theory and statistics.

Geometry (grades 10 and 11) — Euclidean geometry in three-dimensional space: axioms; the relative positions of straight lines and planes; polyhedra and circular bodies; areas of surfaces and volumes; vectors and coordinates in three-dimensional space; introduction to non-Euclidean geometry.

Mathematics (grades 10 and 11) — the contents of this course corresponds in volume to the contents of the courses in algebra and elementary calculus and geometry (united into one) for grades 10 and 11, but is less thorough in character.

Specialists who are qualified as "teachers of mathematics" in Russia are prepared to work in basic and senior secondary general education schools. They can also teach mathematics in secondary professional educational institutions, and they can work within the system of supplementary mathematics education.

3 Contemporary Trends in the Development of School Mathematics Education

The trends in the development of school mathematics education in Russia today are largely determined by the transformations being implemented in the system of general education as a whole. Among the latter we should single out a competence-based approach to education; the humanization, democratization, and humanitarization of the educational system; the introduction of modern educational techniques (including the use of computers as instruments and means of education) (Education, 2004).

A competence-based approach to teaching students in general education schools involves, first and foremost, a change in the approach toward the results of teaching in general education schools (Khutorskoy, 1998). Instead of a body of assimilated information, manifested in the knowledge, skills, and abilities that are formed during the study of each school subject, the expected result of education is now the students' ability to act in a situation of uncertainty. This ability is referred to as the students' "competence." Various groups of "competences" are identified. The basic aim of school education is declared to be the formation of key competences (Lebedev, 2004), which are formed in the process of solving problems that are by and large multidisciplinary, requiring the use of knowledge from different subject areas.

The implementation of a competence-based approach for contemporary school mathematics education signifies increased attention toward practical (applied) problems, as well as problems that require the establishment and utilization of connections between mathematics and other subjects. In addition, a greater role must be played by problems whose solutions are obtained by methods not already known to the students, and, hence, whose solutions presuppose, above all, a search for such methods.

"The humanization of education involves orienting the educational process around the development and self-realization of the individual, around a focus on universal human values, around the optimization of the interactions between the individual and society. The humanization of education is aimed at creating forms, contents, and methods of education and training that will provide for an effective manifestation of a student's individuality — his or her cognitive interests, personal qualities..." (Vishniakova, 1999, p. 61). The humanization of general secondary education involves focusing on methods of teaching aimed at students' development and creating conditions for their self-realization. Such methods of teaching have found expression in the so-called system of personality-oriented education (Serikov, 1999).

As for the humanization of mathematics education, it is pursued through the construction of modern teaching methodologies, aimed at the development of, for example, different qualities of knowledge (awareness, dynamism, etc.) and various forms of thinking (critical, spatial, probabilistic, etc.) that employ mathematical means (Stefanova and Podhodova, 2005). The humanization of mathematics education also is reflected in the implementation of a specialization at the higher stages of education, as described above, since the choice of an educational program with an appropriate profile (and the corresponding course in mathematics) allows students to fulfill their individual educational needs and interests.

The democratization of general education in Russia aims at providing every person — no matter where he or she may reside, no matter what educational institution he or she may attend — with access to quality education at any level (secondary, professional, post-graduate). To this end, a system of distance learning in mathematics is currently being developed. In addition, standardized graduation exams have been introduced in all schools in the country (they are referred to as "uniform state exams"). The results of these exams are screened when students are admitted to institutions of higher learning. As of 2009, the uniform state exam in mathematics is mandatory, as is the uniform state exam in the Russian language.

Lastly, the humanitarization of education, which is conceived of as an "emphasis on the humanities in the educational process" (Vishniakova, 1999, p. 62), is pursued in mathematics education through the inclusion into mathematics education of components that are ordinarily studied in the humanities: historical facts, linguistic questions, facts about the use of mathematical knowledge in everyday life, and other areas of knowledge. The humanitarization of mathematics education may also involve introducing into the teaching process methods for assimilating mathematical knowledge that are more typical of the humanities.

4 The History of the Formation of the System of Mathematics Teacher Preparation in Russia

The state system of mathematics teacher preparation appeared in Russia during the reign of Peter I at the beginning of the 18th century. The transformations that were being carried out in Russia during that era, in connection with the building of the fleet and the development of trade, called for specialists capable of reading diagrams and making calculations. To this end, so-called "cipher" or arithmetic schools were opened all over Russia. The "cipher school" was a type of elementary school (Ovsyankin, 2000). Higher levels of education were offered to students at the artillery and naval schools. It was the students who graduated from these schools who became the first teachers of mathematics and the natural sciences (first and foremost, geography) in "cipher schools." The first Russian mathematics textbook, L. Magnitsky's *Arithmetic*, was written at this time.

In 1779, the first pedagogical (teachers') seminary in Russia — called a "baccalaureate institute" at the time — was founded as an affiliate of the gymnasium of Moscow University. Incidentally, the baccalaureate degree, which was awarded to the graduates of this institute, was an academic degree of a pedagogical character. Soon, another teachers' seminary appeared in St. Petersburg. Besides teaching general subjects, these seminaries devoted a great deal of attention to teaching future teachers the "methods of teaching," or as we would now put it, teaching methodology.

But graduates from these two educational institutions could not fulfill the need for qualified teachers across the enormous expanse of Russia. For this reason, in the 19th century the responsibility for teacher preparation, including mathematics teacher preparation, shifted to the universities.

The preparation of mathematics teachers in Russia during the 19th century took place in physics and mathematics departments at the universities, as well as in three- and four-year teachers' seminaries and institutes (Andronov, 1968). At the universities, teachers were prepared for work in gymnasiums, where students received a thorough and varied education, which they could then continue in institutions of higher learning (mainly universities) after graduating. Teachers' seminaries and institutes prepared teachers of mathematics for so-called "chief schools" in the capitals of the provinces.

At the universities, students received a sound mathematical education, studying so-called "higher mathematics" (number theory, advanced algebra, calculus, advanced geometry, theory of differential equations, theory of probability, etc.). What was lacking at the universities, however, was preparation in the pedagogical disciplines, above all in the methodology of teaching mathematics to students of different ages. In addition, the content of "elementary mathematics" (the set of mathematical facts and problem solving methods covered in school) was not examined. It was expected that the graduates of the universities would develop all of the requisite professional skills in the course of their teaching activity.

By contrast, students at the teachers' seminaries and institutes did not study higher mathematics. Instead, they examined an expanded and elaborated version of the gymnasium course in mathematics. In doing so, they also studied pedagogy, its history, and the methodology of the subject (mathematics) taught in schools. Along with receiving a theoretical education, students acquired experience conducting classes at model schools organized at their seminaries or institutes.

Each of these systems of mathematics teacher preparation had its pluses and minuses. Thus, graduates from the universities had a sound scientific background in mathematics at the expense of pedagogical training. Graduates from teachers' seminaries and institutes, on the other hand, had sufficient professional preparation in pedagogy, but no education in higher mathematics. This led to a recognition of the need to change the content and process of mathematics teacher preparation.

The preparation and implementation of a new system of mathematics teacher preparation, which was developed at the beginning of the 20th century, involved the participation of famous Russian mathematicians like the academicians M. V. Ostrogradsky (1801–1861), V. Ya. Bunyakovsky (1804–1889), O. S. Somov (1815–1876). They participated in the writing of new mathematics curricula, methodological guidelines, and textbooks. Pedagogues-mathematicians who made an enormous contribution to the formation of the modern system of mathematics teacher preparation in Russia also should be mentioned. These included A. N. Strannolyubsky (1839–1903), who was the young Sofia Kovalevskaya's mathematics teacher. Kovalevskaya was the first woman in Russia to become a professor of mathematics. Strannolyubsky was the author and creator of Russia's first algebra methodology manual, "A Course in Algebra Based on the Gradual Generalization of Problems in Arithmetic: For Teachers" (1863). He taught the first higher courses for women in St. Petersburg, in which female mathematics teachers were prepared for so-called higher elementary schools (which offered seven years of studies) and the lower grades of gymnasiums. Another famous pedagogue-mathematician of the beginning of the 20th century was A. N. Ostrogorsky (1840–1917). Ostrogorsky was the author of Russia's first manual on the methodology of geometry, "Materials on the Methodology of Geometry" (1884). He was also the founder of a pedagogical journal, *The Pedagogical Digest* [*Pedagogichesky sbornik*].

The outcome of this transformative activity was the creation at the beginning of the 20th century of institutions of higher learning such as the *Pedagogical Academy of the Education League*, as well as the appearance of pedagogical courses in several school districts (St. Petersburg, Moscow, Kiev, Kazan). These courses prepared university graduates for teaching in secondary schools.

After the Revolution of October 1917, all secondary schools in Russia were transformed into a unified system of "labor schools." The number of schools increased substantially. There was a need for many qualified teachers, including teachers of mathematics. They began to be prepared at pedagogy departments in universities and at specially organized pedagogical institutes, the most prominent among which were the Herzen Pedagogical Institute in Leningrad and the Lenin Pedagogical Institute in Moscow. Teachers' institutes were founded to provide accelerated (two-year) teacher preparation. By 1956, they had practically ceased to exist due to the weak subject (mathematics) preparation they offered to their students.

5 The Traditional System of Mathematics Teacher Preparation in Russia

Teachers of mathematics in Russia (and in the Soviet Union) received their preparation at pedagogical institutes, many of which are now called pedagogical universities.[3] The largest pedagogical universities today are Moscow State Pedagogical University (MPGU) and Herzen State Pedagogical University of Russia (RGPU) in St. Petersburg.

Mathematics teacher preparation traditionally focused on either one field of expertise (mathematics) or two (for example, mathematics and physics, mathematics and informatics). Preparation in one field usually lasted four years, preparation in two fields lasted five years. Students who had completed their studies with passing grades in their classes and who had passed their state evaluation received accreditation as teachers of mathematics or as teachers of mathematics and physics, or mathematics and informatics, respectively. To pass the state evaluation, students had to pass one or several government exams and to defend a final thesis.

Mathematics teacher preparation took place and continues to take place through both on-site and correspondence education. In correspondence education, teacher preparation lasts one year longer than in on-site education.

Over the past eight years, in keeping with state requirements (Government Education Standards in Special Fields, 2000), teacher preparation in a single field takes place over a five-year period. Starting in their third year of studies, students begin to specialize in a second subject area of their choice (for example, informatics, applied mathematics, economics). In such cases graduates cannot obtain accreditation

[3]All educational institutions in Russia that are involved in the professional preparation of licensed experts are commonly called "institutions of higher learning." Below, we will use the term "pedagogical institute" to denote any pedagogical institute of higher learning or university.

as teachers in this subject area (for example, as teachers of informatics). They receive accreditation only in their major field of study.

The program of preparing teachers in different fields includes the following sequences of subjects:

(1) general humanitarian and socioeconomic subjects (philosophy, Russian history, Russian language and culture, foreign language, physical culture, economics, law, etc.)
(2) general mathematical and natural scientific subjects (physics, mathematics (introduction), informatics, ecology, etc.)
(3) general professional subjects (pedagogy, psychology, mathematics education methodology, etc.)
(4) specialized subjects (algebra and number theory, geometry, calculus, differential equations, mathematical logic, elementary mathematics, etc.)
(5) subjects in the students' specializations.

In addition to mandatory subjects, where the sequence and duration of study are determined by the educational plan, the program includes elective subjects that students are free to choose on their own. About 15% of the total time of study is allocated for these elective subjects. Every school year, a new set of elective subjects is offered. The offerings are determined by instructors' interests, changes in the system of general secondary education, and students' needs. The sequence of these subjects and the time spent studying them are determined by the general logic of the program of study, above all by the students' readiness to comprehend the information being offered to them.

Students devote their first three years mainly to studying subjects in the first three sequences. Particular emphasis is given to the specialized (subject) and psychological–pedagogical preparation of future teachers of mathematics. Beginning with the third year of studies, greater attention begins to be devoted to the future teacher's specialized professional preparation, which is sometimes referred to as methodological preparation.

Methodological preparation is that part of a student's preparation which ensures his or her readiness to fulfill the functions of a mathematics teacher at a general education school. It includes theoretical preparation in the following mandatory subjects: school-level mathematics and the theory and methodology of mathematics education. In addition, each student must choose among several elective courses with a psychological–pedagogical (two courses to be selected) and a methodological (three courses to be selected) content. Along with theoretical preparation, the students undergo practical preparation: pedagogical practical training in schools over a 20-week period.

The content of the state evaluation is determined by each pedagogical institution separately. For example, graduates of the mathematics department of Herzen State Pedagogical University of Russia, who have completed the program of study described above, must pass a state exam that contains three questions. The first question requires students to demonstrate knowledge of the mathematics that they have covered in the specialized subject sequence (sometimes referred to as higher mathematics). The second question pertains to school-level mathematics. In answering it, students must demonstrate an understanding of the way in which scientific mathematical knowledge is transformed into the educational knowledge that is studied at the school level.

Finally, the third question requires students to demonstrate an ability to select and arrange teaching materials and methods of teaching in order to provide for the education of schoolchildren of a given age. Most often, the students are asked to work out a methodology for studying a specific topic in the school course in mathematics. It should be noted that the last question is given to students three days prior to the exam. On the day of the exam, before the exam begins, they present the plan which they have developed in a preliminary fashion to the members of the state evaluation committee. During the exam, the students present the results of their methodological schema orally in summary fashion.

Below is an example of the types of questions that students must answer on state exams:

1. Bases of vector spaces. Dimension. Examples.
2. Using the properties of functions and reading graphs in solving equations and inequalities.

3. Develop a set of assignments for studying the concept of the "derivative" at the reinforcement-of-new-materials stage. Describe a methodology for using these assignments so as to facilitate differentiation in education.

The topic of the final thesis depends on the department in which the thesis is written. Students may write their theses in the specialized subject (mathematics) department, in the department of mathematics education methodology, or in the pedagogy or psychology departments. The topic and content of theses written in the pedagogy or psychology departments must be connected with the process of teaching mathematics.

6 Requirements for the Mathematics Teacher in Russia Today

School teachers in Russia, including teachers of mathematics, have always been responsible for the quality and level of instruction offered to the schoolchildren entrusted to them. Today, too, a teacher's work is judged above all by how well his or her pupils know the subject. But the requirements that mathematics teachers in general education schools are presented with today are being augmented as a result of the trends in the development of school mathematics education.

All requirements for teachers of mathematics are based on the following principle: mathematics teachers must know their subject well and be able to teach it to their students.

Let us take a closer look at this principle. What is meant by "knowing a subject well": knowing the educational content and methods for solving the problems found in school textbooks, or understanding the methodology of mathematics as a science, grasping the basic facts of this science and the methods for obtaining them, and grasping the directions in which mathematics may develop in the future? First and foremost, "knowing a subject well" means having knowledge of the content of the school mathematics curriculum and being able to solve school-level mathematical problems. Moreover, the ability to solve mathematics problems and the ability to teach students how to do this is emphasized above everything else. After all, it is recognized universally that the process of searching for and carrying out the solution to a mathematics problem possesses a significant developmental effect and familiarizes students with mathematical activity.

What, then, are the mathematical problems that teachers must be able to work with? First and foremost, these are the problems that are found in school textbooks, problems that appear on the different versions of the uniform state exam, problems that are given to students in various subject competitions (contests, olympiads).

To convey their level of difficulty, several examples of such problems are given below:

1. Put the following numbers in increasing order: $\arcsin\frac{\pi}{6}$; $\arcsin(-0,3)$; $\arcsin 0,9$ (Kolmogorov, 2007, p. 69, problem 134(a)).

2.1. [4]Solve the inequality: $\frac{6x+18}{7x} \leq 0$.
Multiple choice answers: (1) $[-3;0) \cup (0;+\infty)$; (2) $[-3;0)$; (3) $[-3;+\infty)$; (4) $(-\infty;-3] \cup (0;+\infty]$.

2.2. Calculate the value of the following expression: $6^{\log_6 5} + 100^{\lg\sqrt{8}}$.

2.3. Given a regular pyramid *FABC* circumscribed by a sphere with the center in the plane *ABC*. A point *M* lies on the edge *AB* in such a way that $AM : MB = 1 : 3$. A point *T* lies on the straight line $\overleftrightarrow{AF}$ and is equidistant from points *M* and *B*. The volume of the pyramid *TBCM* is equal to $\frac{5}{64}$. Find the radius of the sphere circumscribed about the pyramid *FADC*.

3. For what real values of x and y can one find a number a such that the following inequalities are satisfied: $a < x < a^4 < y < a^2$?
(A) $x = 0$, $y = 1$; (B) $x = -1$, $y = \frac{1}{2}$; (C) $x = 0$, $y = \frac{1}{2}$; (D) $x = -1$, $y = 1$; (E) $x = 1$, $y = 2$. (Plotkin, 2008).

[4]These three problems are taken from the three main parts of the uniform state exam for 2009 (sample version). The problems in the first part have multiple choice answers. In the problems in the second part, students are required to write down the answer (usually a number) in a blank space. In the problems in the third part, students must provide a full solution.

Along with such knowledge of the subject, mathematics teachers today are required to possess knowledge that enables them to establish connections between scientific knowledge and the educational content found in school textbooks. In addition, mathematics teachers are now required to see the connections and the possibilities for using mathematics in other sciences and subject areas (natural sciences and humanities). And this means that teachers must be very well educated not only in their own subject area, but also in other subject areas, and have a high level of cultural literacy.

The aforementioned requirements stem from the fact that the teacher of mathematics is charged with the responsibility to teach not only mandatory mathematics courses, but also elective courses, which are part of the educational program at the specialized level (in the upper grades). These courses must either deepen the content of the mandatory courses, or expand this content (i.e., they must be built on content that is not part of the curriculum), or shed light on questions connected with the use of mathematical knowledge. Meeting such objectives requires very deep and varied knowledge in the field of mathematics.

Teachers must also possess a command of their subject in order to motivate students of mathematics. This is one of the most important pedagogical problems confronting teachers of mathematics today.

Lastly, the need to provide guidance for students engaged in projects and research — which today represent mandatory components of students' cognitive-educational activities, particularly in the upper grades — also requires teachers to have a good command of mathematical knowledge.

Let us now turn to the second part of the principle formulated above. Teachers of mathematics must know how to teach mathematics to their students. Does teaching mean imparting to the students knowledge of the rules for carrying out various mathematical operations, knowledge that will enable them correctly to reproduce the definitions of concepts and mathematical facts? Or does teaching mean getting the students to understand how to acquire and to apply mathematical knowledge, getting them to recognize the connections between different mathematical facts? Today, teachers of mathematics are required to ensure that their students assimilate mathematics at a level that presupposes an understanding of the essence of mathematical objects, an understanding of the methods for obtaining and transforming them, an understanding of the interconnections between various mathematical concepts and facts. It is precisely such a deep assimilation of the content of the course in mathematics by the students that will allow them to employ mathematical knowledge in different situations, which is required of school graduates today.

In order to ensure that all students understand the educational material, teachers today must be aware of the distinctive psychological characteristics and mechanisms for absorbing new knowledge that different students bring to bear on the assimilation of the content being offered to them (Stefanova and Podhodova, 2005). This awareness must form the foundation on which suitable teaching instruments must then be developed (systems of problems, methods of visual representation, instruments for assessing and correcting knowledge).

In other words, the modern mathematics teacher not only must play the part of a qualified interpreter and supervisor for the students as they go through the process of assimilating mathematical knowledge. He or she also must be a researcher and constructor of an effective plan for assimilating mathematical knowledge.

7 The New Model of Mathematics Teacher Preparation in Pedagogical Institutes

In response to the rising demands made on mathematics teachers, the Russian system of higher professional education is undergoing a transition to a two-tier model (baccalaureate program and master's program). This model of higher professional pedagogical (and not only pedagogical) education is referred to in official documents as "specialized preparation." The system of pedagogical education in Russia today distinguishes between seven different specializations: natural scientific education, physical–mathematical education, philological education, social–economic education, technological education, pedagogy, artistic education. Each specialization encompasses several subject areas. Mathematics teachers are prepared within the framework of a physical–mathematical specialization, which includes four subject areas: mathematics, informatics, physics, and astronomy.

At the first tier of the model being examined here (baccalaureate, four years), mathematics teachers are prepared for the basic school (grades 5–9). At the second tier, over the next two years, mathematics teachers are prepared for the high (specialized) school. At this stage, mathematics teacher preparation takes place through special programs, which are developed at each institution of higher learning and then approved by the Ministry of Education and Science of the Russian Federation.

The new model of teacher preparation in pedagogical institutes contains three stages, each of which lasts two years and has its own goal (Diagram 3). The first two stages take place within the baccalaureate program, the last stage within the master's program.

The first stage may be characterized as the stage of general preparation. At this stage, students study subjects that represent all

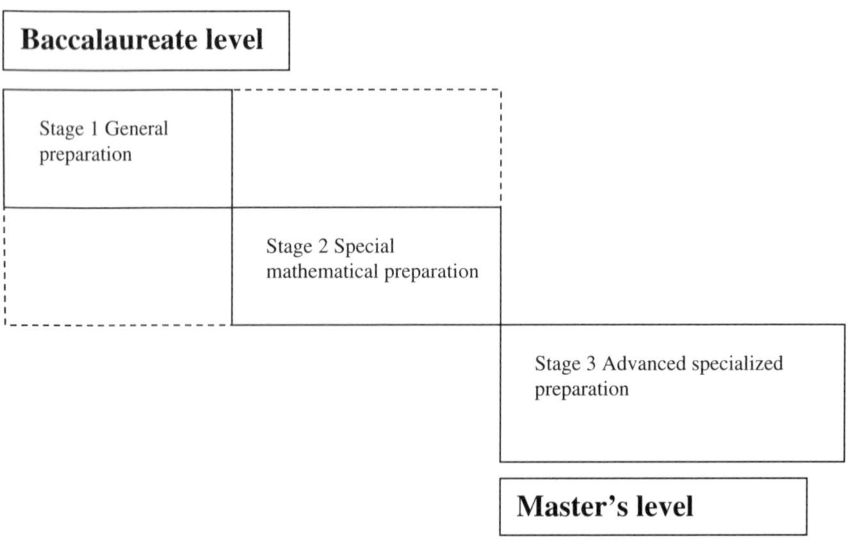

Diagram 3.

fields within a given specialization and select one field for subsequent study.

The second stage is devoted to preparing students in the field which they have selected, as well as to providing them with professional teacher preparation (in the case of future teachers of mathematics, this involves preparation for the basic school, i.e., grades 5–9).

At the third stage, preparation in a specific field continues at a higher level, now with a certain degree of professional specialization.

Before turning to the contents of the professional preparation itself, let us give a general description of each of the aforementioned programs in the two-tier model.

The program which prepares future teachers of mathematics at the baccalaureate level focuses on one of the specializations of higher pedagogical education, namely, physics–mathematics education, within which the program further focuses on a narrower field, mathematics (Government Education Standards in Special Fields, 2000). Students who enter this program study four different sequences of subjects: general humanitarian and social–economic subjects, general mathematical and natural scientific subjects, general professional subjects within their specialization, and subjects of preparation in their specialization. The content of the first two sequences of subjects virtually coincides with the content of the first two sequences of subjects in the specialized preparation program discussed above (general humanitarian and socioeconomic subjects and general mathematical and natural scientific subjects).

The sequence of general professional subjects in the students' specialization includes groups of subjects that represent pedagogy and psychology, the subjects "Technologies and Methodologies of Mathematics Education," "Research in the Field of Physics–Mathematics Education," "Mathematical Models, Methods, and Theories," "Foundations of Discrete Mathematics," "Information Technology in Physics–Mathematics Education," and a number of others.

The sequence of subjects of preparation in specialization "mathematics" includes: "Algebra and Number Theory," "Geometry," "Calculus," "Mathematical Logic and Theory of Algorithms," and "Workshop on Solving Mathematical Problems."

A feature of the baccalaureate preparation program that distinguishes it from the program for preparing specialists is that students in the former become acquainted with the foundations of conducting research in the sphere of physics–mathematics education (see above, "Research in the Field of Physics–Mathematics Education"). Students apply their newly acquired knowledge during practical training in schools, which lasts for eight weeks. Two of these eight weeks are devoted to so-called *educational* training, six weeks to *productive* (pedagogical) training (the first one is mainly devoted to the observation of lessons while the second one includes actual independent teaching).

Upon completing the baccalaureate program, students take a state exam in mathematics and defend a final thesis.

Students who have successfully completed a baccalaureate program in a given specialization may continue their education in a master's program in the same specialization. The master's programs are usually attached to different departments.

The following are the general aims of a master's program in mathematics education:

- to deepen students' knowledge of their subject and professional fields,
- to teach them how to conduct scientific research in their chosen professional field,
- to offer them professional training as teachers of mathematics.

During their studies in the master's program, students go through two types of practical training: training in conducting scientific research (five weeks) and pedagogical training (six weeks). Training in conducting scientific research is aimed at collecting materials for a master's thesis, which students defend as part of their state evaluation. Pedagogical training is aimed at developing a practical ability to implement mathematics education in higher general education schools, in secondary professional educational institutions, or in institutions of higher learning.

As an example, let us look at two specialized preparation master's programs at the mathematics department of Herzen State Pedagogical University of Russia. These programs provide preparation in the specialization of physics and mathematics. The name of the first program is "Mathematics Education," the name of the second is "Mathematics Education in the System of Specialized Preparation."

- The first program is offered at the department of mathematical analysis. It offers in-depth mathematics preparation with a view to preparing graduates for working as teachers of mathematics in institutions of higher learning. Due to this aim, the content of a master's thesis in this program must be based not only on research in the field of mathematics education, but also on solving a research problem in the field of mathematics.
- The second program is offered at the department of mathematics education methodology. The content, research problems, and professional training that it provides are focused on solving problems of mathematics education in high (specialized) schools. This also determines the topics of the master's theses.

Master's programs are seen not only as a particular stage of higher professional education and the professional preparation of specialists, but also as a training ground for acquiring research skills and for selecting the most talented young people for subsequent doctoral studies in graduate school.

In summary, mathematics teacher preparation within the framework of the new model — which today in Russia is becoming established as the basic model — possesses a number of distinctive characteristics. The most salient among them are:

- the greater importance attached to general educational goals over professional goals in the course of teacher preparation;
- the inclusion of a research component in the content of the preparation, which becomes increasingly prominent from the baccalaureate stage to the master's stage;

- the distinction between the professional accreditation of graduates of baccalaureate and master's programs.

8 Selecting Future Students

In entering a mathematics teacher preparation program (or more precisely, in entering a mathematics or physics–mathematics department at an institution of higher learning to which such programs are typically attached), every applicant must go through a competitive selection process to obtain one of a limited number of available places in the program that are paid for by an agency of the federal government, currently the Ministry of Education and Science. The competitive selection process is based on the overall scores received by applicants on their entrance exams.

The number, content, and organization of the entrance exams are determined individually by each institution of higher learning and announced each year in its admission guidelines. The admission guidelines for a given institute are published on the institute's website and in a pamphlet at least six months before the entrance exams are conducted. Entrance exams typically have been conducted during the month of July in written or oral form.

For a long time, applicants took the following entrance exams in order to enroll in a mathematics teacher preparation program: mathematics (written), mathematics (oral), physics (oral), Russian language and literature (written composition).

In recent years, two tendencies have developed in the policy of conducting entrance exams at pedagogical institutes: (1) a reduction in the number of entrance exams; (2) an increased emphasis on written exams over oral ones. In addition, in the assessment of students' scores, students' grade point averages from their school studies (the so-called "average grade of the secondary education certificate") sometimes have been added to their entrance exam scores.

Throughout all of these changes, the written entrance exam in mathematics has remained mandatory for enrollment in a mathematics teacher preparation program. The content of this exam usually includes problems that test students' knowledge of algebra, geometry, and calculus at the school level.

As an example, consider the problems that appeared on the entrance exam given to applicants to the mathematics department of Herzen State Pedagogical University of Russia in 2007. Students were given 4 hours (240 minutes) to complete the exam.

1. Solve the inequality $\frac{1}{x+1} \leq \frac{3}{x+5}$.
2. Find a solution to the equation $\sin 2x - 2\sin x = 0$ on the interval $[-\pi; 0]$.
3. The lengths of the sides of a triangle are known: BC = 4 cm, AC = 5 cm, AB = 6 cm. Find the length of the median drawn to side AC.
4. Solve the equation $2\sqrt{x-2} - \sqrt[4]{x-2} = 15$.
5. Find the domain of the function $y = \sqrt[4]{4 - 3^x - 3^{|x|}}$.
6. Find $\sin^3 \alpha - \cos^3 x$, if $\sin \alpha - \cos \alpha = b$.
7. How many grams of an 8% acid solution can be obtained from 200 grams of a solution that contains 62% sulfuric acid?
8. For which values of the parameter n are the roots of the equation $(1-2n)x^2 = 1 + n(1-3x)$ different and positive?
9. The base of a pyramid is a square. Of two opposite side edges, one is perpendicular to the plane of the base, and the other makes an angle β with the base and has length l. Determine the lengths of the other side edges and the angles of their inclination to the plane of the base.

For the first time, admission to Russian pedagogical institutes in 2009 will be based on students' performance on uniform state exams. In addition to exams that are mandatory for everyone (in mathematics and the Russian language), students will also have to take three exams of their choice. This choice will be based in part on the institute which a student wishes to attend.

Each institute determines different sets of uniform state exams whose results will be considered during the admissions process for students applying to different departments of the institute. For example, in order to enroll at the mathematics department of the Herzen State Pedagogical University of Russia, applicants must present the results of their uniform state exams in mathematics, Russian language, and informatics.

9 The Content and Process of Academic Professional Mathematics Teacher Preparation at the Baccalaureate Level

The professional preparation[5] of the mathematics teacher at the baccalaureate level with a specialization in physics–mathematics education is divided into (1) academic preparation, which takes place at an institution of higher learning, and (2) practical preparation in the form of practical training in school. Consider the first part of this preparation.

The professional preparation of the teacher focuses specifically on the study of such subjects as "Pedagogy," "Psychology," and a set of subjects that represent its methodological component. Among the latter are the "Workshop in Mathematical Problem Solving," "Technologies and Methods of Mathematics Education," and "Research in the Field of Physics–Mathematics Education."

The organization of that part of the professional teacher preparation program which pertains to the teacher's methodological preparation can be examined more closely.

The program of methodological preparation at the baccalaureate level is aimed at imparting the following professional abilities to the students:

- The ability to interpret and to adapt substantively scientific knowledge for the purpose of solving educational problems in the field of mathematics at the levels of the basic school and the senior (specialized) school.

[5]We are emphasizing specifically the teacher's professional (pedagogical and methodological) preparation. It should be underscored once more, however, that a great deal of attention is paid to other aspects of preparation as well. For example, approximately 1500 clock hours are allocated for purely mathematical courses and seminars in the program of preparation at the baccalaureate level.

- The ability to interpret the contents of college-level subject courses (basic courses and electives) in order to enrich the contents of basic, specialized, and elective courses in mathematics.
- The ability to use standard problem solving methods in accordance with the mathematics curriculum of basic general education schools.
- The ability to evaluate the quality of educational literature on a subject for basic schools.
- The ability to use basic technological schemas for working with elements of mathematics (concepts, rules, theorems, problems) when teaching students.
- The ability to construct and implement a mathematics education process (system of lessons) in a basic school while using the most effective teaching technologies.
- The ability to use standard computer programs to facilitate certain types of mathematical and teaching activities.

Note that three educational subjects ("Workshop in Mathematical Problem Solving," "Technologies and Methods of Mathematics Education," and "Research in the Field of Physics-Mathematics Education") that comprise the methodological component of the preparation are aimed at developing specific professional skills and are expected to develop the future mathematics teacher's professional activities in three directions: with respect to subject matter, with respect to organization and methodology, and with respect to research. The aim of the first subject ("Workshop in Mathematical Problem Solving") is for students to assimilate school mathematics professionally through the process of problem solving. The aim of the second subject ("Technologies and Methods of Mathematics Education") is for students to assimilate modern technologies of teaching mathematics in school. The aim of the third subject ("Research in the Field of Physics-Mathematics Education") is to acquaint students with the foundations of and methods for conducting research in the field of mathematics education. All of the subjects are oriented toward future professional work in the basic school (grades 5–9).

Let us now describe the content and organization of education in each of the three aforementioned subjects. As an example, consider materials created and used by the mathematics department of Herzen State Pedagogical University of Russia.

The subject "Workshop in Mathematical Problem Solving" is studied for three semesters (during the third and fourth years of study). Its purpose is for students to acquire professional experience in solving problems in elementary (school-level) mathematics. The study of this subject is aimed at:

- developing the ability to use various methods for solving mathematics problems;
- developing general mathematical and methodological literacy in working with problems.

The basic concepts examined in this course are: the mathematics problem; the word problem; typology of problems; the means, the method, and the technique of solving a mathematics problem; stages in the solution of a mathematics problem. In terms of the content of the course, the following topics may be identified: word problems; divisibility problems; problems on algebraic and transcendental functions, as well as corresponding equations and inequalities; geometric problems (mainly in plane geometry) involving computations, proofs, and constructions, including geometric transformations; logical problems and puzzles.

Thus, the content of the problem solving workshop is aimed principally at developing a deep ability to solve problems that correspond to the contents of the school course in mathematics at the basic school level (grades 5–9). However, the level of difficulty of these problems is considerably higher than the level of difficulty of the problems found in school textbooks. This is due to the fact that these problems and the process of solving them must be interesting to the University students. At the same time, being able to solve more difficult problems meets the requirement of the advanced (professional) level of solving school-level mathematical problems.

The following divisibility problem will serve as an example to demonstrate the level of difficulty of the problems solved in the workshop (Vavilov, 1987, p. 18).

A natural number $n > 1$ is not divisible by 2 or by 3 without a remainder.

Prove that the number $n^2 - 1$ is divisible by 24.

The workshop consists exclusively of practice sessions in which students examine the basic methods for solving problems and analyze the solutions of problems that have given them difficulty. In addition, students are given sets of problems to solve on their own, outside of class. Furthermore, while doing their independent work, students complete assignments aimed at identifying possible difficulties in solving problems of various types, and at determining the basic strategies of searching for the solutions to problems.

The aim of studying the subject "Technologies and Methodologies of Mathematics Education" — which, like the workshop, is offered over a period of three semesters — is to develop a system of knowledge about methods of structuring the process of mathematics education for different categories of schoolchildren and to develop basic professional skills in facilitating the education of students at the basic general education school level.

The content of this subject may be divided nominally into two parts: a fundamental part (oriented toward theory) and an applied part (oriented toward practice). The first part explores such questions as the role and place of mathematics education, the organization of the system of mathematics education in Russia, the components of the methodological system of mathematics education, the interconnection between technology and the methodology of mathematics education, the organization of the content of mathematics education; methodological schemas for teaching students various components of the mathematics curriculum; traditional and innovative technologies in the process of mathematics education. The second part explores issues connected with structuring the mathematics education process (using as examples various systems of mathematics lessons for grades 5 and 6, and algebra and geometry lessons for grades 7–9), as well as issues connected with analyzing this process. The content of the first part is quite broad, encompassing the basic components of the mathematics education system, and not just grades 5–9 (the general education school). The second part, by contrast, is constructed around the example of facilitating the process of mathematics education in the basic general education school (grades 5–9).

The subject is taught through lectures and small group workshops, and students also are required to do a considerable amount of independent work. In the first semester, more time is devoted to lectures than to workshops. In subsequent semesters, lectures play a smaller and smaller role, and the last semester is devoted entirely to workshops. The amount of independent work that students are required to do also increases with every semester.

The lectures examine the most general issues connected with the methodology of teaching mathematics in the basic school. Emphasis is placed on the technological aspect of education, i.e., on the question: "How can the education process be organized?" A more detailed account of the basic contents of the lectures may be found in the study guide for the course (Stefanova and Podhodova, 2005).

The practical workshops are devoted to discussions of assignments that the students have completed beforehand (essentially, methodological problems), which are assigned by the instructor. The following is an example of one such assignment:

> Develop a methodology for working with the following theorem: "If the alternate angles formed by the intersection of two straight lines with a transversal are congruent, then the straight lines are parallel" (Orlov, 2007, p. 179).

In the workshop, before beginning their work, students are asked to answer a series of questions (they are given to them in advance) that highlight the facts that are significant for studying the given topic.

For example, before investigating the properties of parallel lines in a plane, students are asked to answer the following questions:

1. What is the formulation of the axiom of parallel straight lines?
2. Can it be concluded that two straight lines are parallel if, when they are intersected by a third straight line, congruent angles are formed?
3. Given two points in a plane, how many pairs of parallel straight lines can be drawn through these points?

4. What kind of proposition is called an "axiom"?
5. What kind of theorem is called a "converse theorem"?
6. Are the inverse and converse of a given theorem logically equivalent?
7. Explain the method of proof by contradiction. (Orlov, 2007, pp. 176–177)

The subject "The Foundations of Research in Physics–Mathematics Education" is studied for one semester, prior to pedagogical practical training in school. It is aimed at developing a foundation of knowledge about theoretical and experimental methods that are employed in conducting research in the field of mathematics education.

The subject occupies an intermediary position. On the one hand, it must be seen as building a preliminary foundation for future programs of preparation at the master's level, in which greater attention is devoted to students' research activity; on the other hand, it must facilitate the solution of practical problems that the students will confront during their pedagogical practical training and that they will have to solve in writing their final thesis.

In light of this, it is imperative:

- to familiarize the students with the most up-to-date trends in research in the field of physics–mathematics education;
- to give them a basic foundation of knowledge about theoretical research methods in the field of mathematics education;
- to give them a basic ability to carry out a theoretical analysis of a problem (comparative and historical analysis of solving a problem, relying on literary sources);
- to give them an understanding of the specifics of pedagogical experiments in methodological research;
- to give them a basic ability to develop diagnostic experiments, and to analyze and to present their outcomes.

In concluding this description of the professional preparation of the mathematics teacher at the baccalaureate level, the leading role played at this stage of the education process by the teacher should be emphasized. This is due above all to the fact that at the baccalaureate level, the formation of professional skills is just beginning.

And therefore it is precisely the instructor who transmits the basic educational information to the University students; he or she directs the process of its assimilation and controls the level at which it is assimilated. The students' independent work (which occupies about 50% of the time allocated for the study of each subject) is intended to facilitate the formation of basic professional skills, and also involves searching for information according to the professor's instructions.

10 The Content and Process of the Professional Academic Preparation of the Mathematics Teacher at the Master's Level

At the master's level, students receive specialized preparatory education. Programs in specialized preparatory education are developed by each pedagogical institute that has the right (license) to prepare master's level graduates. Such programs are usually aimed at deepening students' education and preparation in their area of specialization (in our case, mathematics), facilitating the professional preparation of mathematics teachers for the upper grades of the general education school or the professional secondary education system ("colleges," lyceums), and creating conditions in which students can acquire experience in conducting research.

Graduates of master's programs must possess the following general professional skills:

- the ability to evaluate the quality of the content of the educational literature on a given subject for the general education school;
- the ability to interpret the contents of the educational program scientifically with a view to establishing the various levels at which it may be presented and planning its presentation at various levels;
- the ability to apply different approaches when teaching different specific topics including constructing specific number sets, defining elementary functions, constructing a system of knowledge about three-dimensional geometric objects, measuring magnitudes, teaching a variety of mathematical methods (vectors, coordinates, and the method of geometric transformations), teaching calculus;

- the ability to evaluate mathematics tests;
- the ability to use problem solving methods to solve the problems found in high-school textbooks, including challenging problems and olympiad-type problems;
- the ability to plan elective courses for pre-specialized preparation (grade 9) and specialized education (grades 10 and 11);
- the ability to choose and use suitable computer programs for solving problems arising in mathematics education.

Students are admitted to master's programs on the basis of entrance exams, which usually take the form of an interview. It is expected that the pool of applicants to a master's program will be drawn mainly from the graduates of baccalaureate programs specializing in the same subject (physics–mathematics education). This does not have to be the case. Graduates of baccalaureate programs specializing in any subject can enter a master's program specializing in mathematics education. In such cases, however, in addition to going through an interview, applicants must pass an exam in mathematics whose content is analogous to the content of the exam taken by graduates of baccalaureate programs specializing in mathematics education.

During the interview, students are asked questions of a general nature, which are intended to reveal not only their knowledge of the foundations of mathematics education methodology, but also their critical abilities and their individual professional–pedagogical stance toward the current state of affairs in the modern system of general secondary education.

The following is an example of a topic that might be discussed during an admission interview.

> The organization of students' independent work in mathematics classes and outside of class: traditional and innovative approaches. Its role in the teaching process and difficulties connected with its facilitation.

The main subject devoted to the professional preparation of mathematics teachers at the master's level is "Methodology of Mathematics Education." Depending on the particular orientation of the master's program, the title of this course may vary. For example, in the master's program in "Mathematics Education in the System of Specialized Preparation," this course has the title "Methodology of Mathematics Education in Specialized Schools."

As pointed out above, in a master's program students are prepared for working as teachers in high school, which today are specialized schools. The subject "mathematics" in a specialized school is represented by a basic, specialized course (actually, two courses: "Algebra and Elementary Calculus" and "Geometry") as well as elective courses. In studying the course in methodology students must grasp the specific features of the implementation of each of these courses and acquire some initial experience in planning the educational content of and teaching methods for each of these courses as well. In addition, if today there already exist textbooks for the basic and specialized courses in mathematics, then the content of the elective courses remains undefined, although examples of such courses are available. One of the aims of teaching students is to teach them how to plan elective courses and to teach them the methods for facilitating them.

Planning is a basic form of professional activity. A course in methodology at the master's level may be aimed at getting students to master this form of activity. Planning is an activity that should be contrasted with analytical and interpretational activities, which students are taught to master professionally at the baccalaureate level.

Another crucial feature of master's level preparation is the aim that an overwhelming majority of new knowledge and skills must be acquired by students in the course of independent work. In every course, including the course in methodology, 75% of the time is allocated for independent work by the students, and only 25% of the time is allocated for in-class work with the professor. The professor's role is not to convey information to the students, but to consult them and correct their understanding of the information which they have found and developed on their own.

Several examples of the way in which master's level students' independent work is organized in the Herzen Pedagogical University course on the "Methodology of Mathematics Education in Specialized Schools" are informative.

In order to organize independent work for the students, several generalized professional methodological problems are isolated and concretized in the form of assignments for students. Among them are the following:

1. Conduct a comparative analysis of the educational aims, content, and preferred teaching and testing methods involved in facilitating the basic, specialized, and elective courses in mathematics. In particular, select a specific topic in these three different courses and conduct the comparative study along the aforementioned lines. Also, conduct the comparative study of systems of mathematics problems on the selected topic in these three different courses.
2. Identify the distinctive features of the methodologies employed in facilitating the basic, specialized, and elective courses in mathematics and demonstrate the methods for facilitating them while studying a specific educational content (lesson, topic, section of a course).
3. Develop a system of testing to be employed while covering a topic in the basic, specialized, or elective courses.
4. Develop instructional guidelines for high-school students' research and independent projects of a mathematical or interdisciplinary content.

An example of one way in which the first generalized professional methodological problem may be concretized is given below. Students are given the following assignment to complete on their own (Stefanova, 2004).

> *Identify the mental and mathematical actions that students must perform in order to solve the following mathematics problems.*
>
> Problem 1 (Karp and Werner, 1999, p. 64, No. 6).
>
> On a number line, mark the following numbers: 3^{-n}, n, $\frac{1}{n^2}$, $(-1)^n\frac{1}{n}$, $1+\frac{1}{n}$ for $n = 1, 2, 3, 4$. Which of the following propositions is true?
>
> (a) $\lim_{n\to\infty} 3^{-n} = 0$; (b) $\lim_{n\to\infty} n = 0$; (c) $\lim_{n\to\infty} \frac{1}{n^2} = 0$; (d) $\lim_{n\to\infty}(-1)^n\frac{1}{n} = 0$; (e) $\lim_{n\to\infty}(1+\frac{1}{n}) = 0$.

Problem 2 (Karp and Werner, 2002, p. 26, No. 3).

Numbers $a_n = 6n (n \in \mathbf{N})$ are given. (a) Find all n for which $a_n > 100$; (b) How many natural numbers n are there such that $a_n \leq 1000$; (c) Let k be a given positive number. Prove that there exists a finite number of natural numbers n such that $a_n \leq k$; (d) Find $\lim_{n\to\infty} a_n$; (e) On a number line, mark the following numbers: $\frac{1}{a_1}, \frac{1}{a_2}, \frac{1}{a_3}$; (f) Find all n such that $\frac{1}{a_m} \leq \frac{1}{1000}$; (g) Find $\lim_{n\to\infty} \frac{1}{a_n}$; (h) Find $\lim_{n\to\infty} (-1)^n \frac{1}{a_n}$.

Problem 3 (Vilenkin, Ivashev-Musatov, and Shvartsburd, 2000, p. 143, No. 324).

Prove that the following sequences converge to 0:

(1) $(\frac{n}{2^n})$; (2) $(\frac{2^n}{n!})$; (3) $(\frac{n^2}{5^n})$; (4) $(\frac{(n!)^2}{(2n)!})$.

It should be noted that students are not informed in advance from which textbooks these mathematics problems are taken. Initially, students must solve each of the given problems and assemble a comparative table specifying the actions that schoolchildren must carry out in analyzing and solving these problems. After this, they must say which of the problems is the most difficult. At this point, they are told what textbooks the given problems were taken from and what level of education in mathematics they presuppose.

After discussing the results of the problem solving, the students, with the professor's help, pose possible methodological problems. For example:

- What mathematical knowledge is it expedient to recall before solving these problems? How can this be done in class?
- What kind of discussion would you initiate after solving each of these problems? Why?
- Formulate several (at least three) questions for schoolchildren aimed at getting them to understand the concept of sequences that converge to 0.
- What difficulties might schoolchildren encounter in solving the given problems?

- On your own, make up several simple and difficult problems aimed at getting schoolchildren to grasp the concept of sequences that converge to 0.

By completing this assignment, the students will be able to feel better and to make sense of the difference between the problems employed in the basic and specialized mathematics courses and, therefore, the difference between the distinctive characters of these courses as a whole.

The following assignment illustrates the solution of the second generalized professional methodological problem.

> *Identify the distinctive features in the construction of the methodology of the study of the topic "Derivatives and Their Use" in the basic and specialized courses in mathematics and show examples of the ways in which they might be employed in the study of this topic.*

The basic question for discussion in class is: In what way and how are the distinctive features of teaching the basic and specialized courses in mathematics expressed when students are being taught the topic of "Derivatives and Their Use"?

In order to take active part in the discussion, students are expected to complete the following supplementary assignments on their own beforehand:

- Compare the theoretical materials presented in *explanatory sections in textbooks* on the given topic (its scope, depth, strictness of exposition).
- Answer the following questions: What *mathematical facts and methods* are examined in the texts of different textbooks? Which of them are *essential (fundamental)*? Which mathematical ideas are elucidated (mentioned explicitly or indirectly) in presentations of this topic? What makes the given educational material a part of the general cultural baggage of modern human beings?
- Identify groups of problems on the given topic that are represented in different textbooks. Evaluate *the level of difficulty of problems* that you group together.

- Identify the main *methodological approaches* to presenting the teaching materials pertaining to this topic which are embraced by the authors of different textbooks.
- Determine which materials pertaining to this topic must be examined in class with the students and which materials must be assigned for independent work at home.
- Determine the basic problems assigned to students in the topic. Suggest techniques for working with them.
- Identify techniques that students may employ in working with the educational materials pertaining to the topic. Provide examples of such techniques.
- Provide examples of problems that may be used for testing how well students have assimilated the materials pertaining to the topic.

In order to successfully complete the assignment, students are expected to make use of high-school textbooks (Bashmakov, 1991; Kolmogorov, 2007; Karp and Werner, 1999; Vilenkin, Ivashev-Musatov, and Shvartzburd, 2000) that are oriented toward different specializations.

In the course "Methodology of Mathematics Education in Specialized Schools," questions about the content and organization of students' activity in studying mathematical elective courses are examined separately. Students are given the opportunity to participate in collective projects (usually in groups of two or three) in which they create a program and a methodology for implementing an original elective course for schoolchildren.

11 The Practical Preparation of the Mathematics Teacher (Field Practice)

We will now turn to the students' professional practical preparation. It is facilitated through a system of pedagogical practical training, which includes, as was mentioned above, *educational* practical training and *productive* practical training (at the baccalaureate level) and *pedagogical* practical training (at the master's level).

Every type of practical training has a predetermined aim. Taken together, the aims of the different types of practical training are determined by the aims of the practical preparation of the teacher as a whole. As an example, look at the aims of the professional practical trainings which were developed at the mathematics department of Herzen State Pedagogical University of Russia.

In the course of their *educational practical training*, students must acquire experience in analyzing an actual mathematics class as well as to conduct a few psychological mini-investigations.

The aim of the *productive practical training* consists in the formation of an ability to plan and facilitate work with individual components of mathematics in mathematics classes in basic schools, and in giving students initial experience in conducting separate lessons in mathematics (algebra and geometry) and working outside of class with students (grades 5–9).

Pedagogical practical training at the master's level is aimed at getting the students to acquire an ability to search for and interpret scientific–methodological materials for the purpose of independently planning for modern mathematics education in a high school, while taking into account the distinctive features of its specialization.

The activity of the students during their practical training is organized by means of specially developed methodological problems.

Two sample problems are given below: the first to illustrate the kinds of problems that students must solve in the course of their educational practical training, the second to illustrate the kinds of problems that students must solve in the course of their productive practical training.

Problem 1. *Forming an individual-personal characterization of the student.*

1. Assemble a psychological portrait of one student at a school. For this assignment, make use of school documentation, observation, questionnaires, conversation, sociometry, etc.
2. Propose practical recommendations for organizing interactions with this student.
3. Formulate a hypothesis about:
   - the influence of interests and hobbies on students' success in mathematics;
   - the relation between general and mathematical literacy;
   - the role of motivation in learning mathematics.

4. Based on your research, assemble a pedagogical character sketch of the student and formulate propositions for planning out his or her individual educational trajectory.

Problem 1 will be considered solved if:

- the teacher-learner has become acquainted with at least the educational program and regulations of an educational institution, the work plan of the homeroom teacher;
- in assembling the pedagogical character sketch of the schoolchild, the teacher-learner has made use of at least two diagnostic methodologies;
- the teacher-learner has been able to formulate in an informed way at least one of the hypotheses in the problem;
- the teacher-learner has described most of the schoolchild's individual-personal characteristics in the pedagogical character sketch of the schoolchild.

Problem 2. *Planning the process of working with word problems.*

1. In conversation with the teacher and the supervisor of the practical training, try to obtain the answers to the following questions:
   - What is the place occupied by word problems in the mathematics education process?
   - What distinctive features can be identified in the way in which students perceive word problems in contrast with the way in which they perceive other types of mathematics problems?
   - What skills are formed at each stage of working with word problems?
   - What stages of working with word problems are the most difficult from the point of view of organizing students' activity?
   - What portion of the students in a class is able to solve word problems successfully?
   - What difficulties do students experience in solving word problems?

2. Find out what word problem will be solved during the next lesson and solve it. While solving the problem (analyzing the problem, searching for its solution, and writing down the solution), formulate the series of questions that shape your own work with the problem. Record these questions and the answers you give, breaking them down into groups in accordance with the stages of your work on the problem (by filling the table with the following rubrics).

Stage of working on problem	Question	Answer

3. Transcribe the stage of the lesson during which the students and teacher were working on the word problem in terms of the following schema:

Teacher	Student	What was written on the board	What students wrote in their notebooks

4. Based on the transcript, identify on your own the different stages of working on the problem.
5. Develop an optimal series of questions supporting work with the given problem.
6. Evaluate the students' involvement at each of the identified stages; determine what techniques are used by the teacher to increase their involvement.

Problem 2 will be considered solved if:

- the teacher-learner has put together a system of questions for the word problem which will be solved in class, in accordance with the stages in which work on the problem will progress;
- the teacher-learner, in accordance with this schema, has transcribed the stage of the lesson during which this problem was solved, identifying the stages in which work on the problem progressed;
- the teacher-learner has been able to demonstrate the optimality of his or her series of questions for working with the given problem.

The master level students' pedagogical practical training includes:

- planning the content and facilitation of a system of interconnected lessons[6] for the basic and specialized courses in mathematics;
- planning the content and organization of the teaching process for an elective course[7] in mathematics;
- developing and receiving approval for innovative instructional–methodological materials (computer presentations, training aids, diagnostic materials, materials for different types of interim testing, materials for organizing independent cognitive-educational activities for the students).

All types of practical training are organized at the university's basic schools. As an exception to this rule, at the concluding stage of study, for students who are already working in secondary general education institutions, practical training can be organized on-site at these institutions.

Every university student is assigned to a professor in the department of the methodology of mathematics education. This professor serves as supervisor for the practical activities of the students in schools. Students participate in *educational* practical trainings in groups of 10–12. Other types of practical training are conducted with groups of 4–8 students per school.

As a rule, a teacher-learner is assigned to one class in which he or she conducts not only teaching work, but also all other types of work with the students typical for a homeroom teacher. The school teacher of this class serves as the student's advisor and consulting expert.

During their practical training the students keep journals, which are considered part of their individual portfolios. In addition, they are given models for formatting the materials which they must develop

---

[6]Over the course of the practical training, the student must analyze and conduct no fewer than nine interconnected classes for the upper grades of the natural scientific-mathematical or humanitarian group of specializations, and no fewer than three interconnected classes for the opposite group of specializations.

[7]The teacher-learner must conduct no fewer than three classes of an elective course.

over the course of their practical training: specific ways of recording and analyzing lessons; schemas for lesson transcripts.

The supervisors of the practical training receive evaluation forms for evaluating the activity of the students in their practical training, with criteria for assessing different aspects of the students' work. The students also are familiarized with these evaluation forms.

Pedagogical practical training is considered an integral component within the professional preparation of the mathematics teacher. On the one hand, it serves as a testing ground for applying knowledge received in the course of theoretical instruction; on the other hand, it constitutes an independent component in the process of educating the students.

12 Problems and Approaches to Solve Them

The problems that arise today in the Russian system of mathematics teacher preparation largely stem from the general changes taking place within the entire system of professional education in Russia. The transition to a two-tier model of higher professional education, traditional in the West, necessitates the construction of a fundamentally new system of teacher preparation.

The traditional system of higher professional pedagogical education which evolved in Russia (and this held for the education of engineers and medical professionals as well) was structured in a way that put the greatest emphasis on receiving certification, which was given to students when they graduated from an institution of higher learning, while the educational program necessary for receiving certification was subordinated to this aim and organized with a view to achieve it. Incidentally, this logic led to the creation of a system of institutions of higher learning in Russia (and the Soviet Union) which consisted largely of "sectorial" (pedagogical, engineering–technological, medical) institutes (now universities). By contrast, the system that is taking shape today puts the greatest emphasis on the problem of obtaining a modern education. This aim has been pursued traditionally in the so-called traditional universities.

Questions arise about the nature of the education that should be offered by institutions of higher learning. For example, should there be differences between the mathematics education of, say, a future research mathematician and a future mathematics teacher? On this score, there are, today, at least two opposing views. The first view is that education should be the same for both. It is better if it is obtained at a traditional university. The second view claims that the education of a future scientist and the education of a future mathematics teacher — even the mathematics education of a future mathematics teacher — must be different. In mathematics courses for future mathematics teachers, different points should be underscored; for example, the connections between "higher" (college-level) and "elementary" (school-level) mathematics should be highlighted. And the content of such professionally-oriented mathematics courses has been developed and improved at Russia's leading pedagogical institutes for a long time.

Another question, which is also connected with the content of the education of the future mathematics teacher, is: what is more important, subject education or psychological–pedagogical education? As in the case of the previous question, two opposite answers are given to this question. The first answer is that, of course, future mathematics teachers need not only profound and professionally oriented mathematical knowledge, but also a broad spectrum of psychological–pedagogical knowledge, that will help them to educate their students more effectively. The second answer is that mathematics teachers (particularly in high schools) need above all a sound education in mathematics, while they can acquire psychological–pedagogical knowledge independently (if necessary) when they are already working as teachers. Without entering into polemics with those who hold the latter view, I should like to note that — although I understand the motivation behind the latter view — I favor the former.

The views being expressed today focus on the problem of the need for a system of pedagogical education in Russia that is independent of the traditional university system. More and more often, one hears explicit and implicit calls for the dismantling of the system of higher pedagogical education and the expansion of the system of traditional university education. In practice, this is reflected in the fact that a number of pedagogical institutes in Russia have been transformed into traditional universities, while others have become part of newly formed universities.

Despite all of the structural changes taking place in the system of higher pedagogical education, it is unlikely to disappear. First of all, because for the Russian system education, this would constitute a step backward. Second of all, the existing and developing state educational standards for higher professional education identify pedagogical education as a separate sector. Furthermore, the standards for higher pedagogical education treat the professional preparation of students (for example, for carrying out the duties of a mathematics teacher) as a problem of equal importance to the problem of obtaining a fundamental and professionally oriented general cultural, mathematical, and psychological–pedagogical education.

Another argument for the preservation of the system of higher pedagogical education is the motivation of the students in the educational process. Observations and opinions of many attest to the fact that in those cases when professionally motivated students — students who have already decided to choose the profession of mathematics teacher in the future — come to an institution of higher learning, they are more successful at assimilating the educational program and the program of professional preparation than students just receiving a general preparation who are uncertain about their future job.

13 Conclusion

The preparation of the mathematics teacher in Russia can be viewed as an evolved system with long historical traditions behind it. It acquired its current form largely during the existence of the Soviet state. It may be divided into academic and practical components. Its academic component includes general cultural, mathematical, and psychological–pedagogical education, as well as the methodological preparation of the future teacher. It is facilitated through academic subjects that students study at institutions of higher learning. Its practical component is facilitated through various forms of practical training in general education schools.

In 10 years since the establishment of government education standards for professional pedagogical education in 2000, teacher preparation has been facilitated within a one-tier framework — the traditional system of mathematics teacher preparation with a specialization in mathematics, and a new two-tier framework of mathematics teacher preparation at the baccalaureate level for the basic school and at the master's level for the high school.

The appearance of a new model of higher pedagogical education stems from Russia's endorsement of the Bologna Declaration, as well as by the new demands being made due to processes taking place within the system of general secondary education, above all, by the introduction of different specializations at the high-school level.

The construction of the system of the professional preparation of mathematics teachers within the new (two-tier) framework of higher pedagogical education is not complete. In many ways, this system relies on traditions that evolved in Russia during earlier historical periods. At the same time, educators are developing fundamentally new approaches both to select the content of education and to enrich educational technologies — approaches that correspond to the new organization of the education process and to the new demands placed on the modern mathematics teacher.

References

Andronov, N. K. (1968). Podgotovka prepodavateley matematiki dlya gimnazii i uchiteley dly narodnykh uchilisch v Rossii i polveka razvitiya novoy sistemy podgotovki matematikov-pedagogov edinoy sovetskoy shkoly SSSR [The preparation of mathematics teachers for gymnasiums and of teachers for public schools in Russia and fifty years in the development of the new system of preparation of mathematicians — pedagogues of the uniform Soviet school in the USSR]. *Uchenye zapiski MOPI*, 202(6). Moscow: MOPI.

Bashmakov, M. I. (1991). *Algebra i nachala analiza. Uchebnik dlya 10-11 klassov sredney shkoly* [*Algebra and Elementary Calculus. Textbook for Grades 10–11*]. Moscow: Prosveschenie.

Bologna (1999). *The Bologna declaration on the European space for higher education. Joint declaration of the European Ministers of Education. 19 July 1999 [Bologna. Zona evropeiskogo vysshego obrazovaniya. Sovmestnoe zayavlenie evropeiskikh ministrov obrazovaniya 19 Iulya 1999]*. [www.spbu.ru/News/edusem/bol.htm].

Education. Conception of modernizing of Russian education till 2010. (2004). [www.edu.rin.ru]

Government Education Standards in Special Fields. Government education standards for higher professional education in the field "mathematics" (qualification: teacher of mathematics [GOS po spetsialnostyam. Gosudarstvennye obrazovatelnye standarty vyshego professionalnogo obrazovaniya po spetsialnosti "matematika" (kvalifikatsia — uchitel matematiki)]. (2000). [www.teacher-edu.ru/wmc/standarts] or [www.informika.ru].

Government Education Standards. Mathematics. The Federal Component of the Government General Education Standards. Mathematics [*Gosudarstvennye obrazovatelnye standarty. Matematika. Federalnyi component gosudarstvennogo standarta obschego obrazovaniya. Matematika*]. (2004). Moscow: MOiN RF [www.ed.gov.ru].

Government Specialized Education Standards. Government education standards for higher professional education/specialization: Physics–mathematics education [GOS po napravleniyam. Gosudarstvennye obrazovatelnye standarty vysshego professionalnogo obrazovaniya/napravlenie — fiziko-matematicheskoe obrazovanie] (2000). [www.teacher-edu.ru/wmc/standarts] or [www.informika.ru].

Karp, A. P., and Werner, A. L. (1999). *Matematika. Uchebnoe posobie dlya 10 klassa gumanitarnogo profilya* [*Mathematics. Textbook for Grade 10 with a Specialization in Humanities*]. Moscow: Prosveschenie.

Karp, A. P., and Werner, A. L (2002). *Matematika. Praktikum po resheniu zadach. Uchebnoe posobie dlya 10 klassa gumanitarnogo profilya* [*Mathematics. Problem Solving Practice. Teaching Manual for Grade 10 with a Specialization in Humanities*]. Moscow: Prosveschenie

Khutorskoy, A. V. (1998). *Evristicheskoe obuchenie: teoriya, metodologiya, praktika* [*Heuristic Education: Theory, Methodology, Practice*]. Moscow: Mezhdsunarodnaya pedagogicheskaya akademiya.

Kolmogorov, A. N., Abramov, A. M., Dudnittsyn, Iu. P. *et al.* (2007). *Algebra i nachala analiza. Uchebnik dlya 10–11 klassov* [*Algebra and Elementary Calculus. Textbook for Grades 10–11*]. Edited by A. N. Kolmogorov. Moscow: Prosveschenie.

Kozyrev, B. A. Radionova, N. F., and Triapitsyna, A. P. (Eds.). (2005). *Kompetentnostnyi podkhod v pedagogicheskom obrazovanii* [*The Competence-Based Approach in Pedagogical Education*]. St. Petersburg: Izdatelstvo RGPU.

Lebedev, O. E. (2004). Kompetentnostnyi podkhod v obrazovanii [The competence-based approach in education]. *Shkolnye tekhnologii*, *5*, 3–12.

The Mathematics Curriculum. Curriculum for General Education Schools, Gymnasiums, Lyceums: Mathematics, Grades 5–11 [*Programma po matematike. Programma dlya obscheobrazovatelnykh shkol, gimnazii, litseev: Matematika. 5–11 klassy*]. (2002). Moscow: Drofa.

Orlov, V. V. (Ed.). (2007). *Metodika i tekhnologiya obucheniya matematike. Laboratornyi praktikum* [*Methodology and Technology of Mathematics Education. Laboratory Workshop*]. Moscow: Drofa.

Ovsyankin, E. G. (2000). *Iz istorii podgotovki rossiiskikh uchiteley [From the History of Teacher Preparation in Russia].* [www.avpu.ru/history].

Plotkin, A. I. (Ed.). (2008). *Vse zadachi "Kenguru"* [*Every "Kangaroo" Problem*]. St. Petersburg: Levsha.

Profiled (2004). Conception of profiled education at Russian high school [www.profile-edu.ru].

Required Content Minimum. Required Content Minimum for Secondary (Complete) General Education. Mathematics [*Obyazatelnyi minimum soderzhaniya. Obyazatelnyi minimum soderzhaniya srednego (polnogo) obschego obrazovaniya. Matematika*]. (1999). Moscow: MOiN RF [www.ed.gov.ru].

Serikov, B. B. (1999). *Obrazovanie i lichnost'. Teoriya i praktika proektirovaniya obrazovaqtelnykh sistem* [*Education and Personality. The Theory and Practice of Planning Educational Systems*]. Moscow: Logos.

Stefanova, N., and Podkhodova, N. (Eds.). (2005). *Metodika i technologiya obucheniya matematike. Kurs lektsii* [*The Methodology and Technology of Mathematics Teaching. A Cource of Lectures*]. Moscow: Drofa.

Stefanova, N. (2004). Graduate students' preparation for teaching mathematics at a profiled high school. *Didactics of Mathematics*, 5, 41–46. Wroclaw.

Vavilov, V. V. (1987). *Zadachi po matematike. Algebra* [*Problems in Mathematics. Algebra*]. Moscow: Nauka.

Vilenkin, N., Ivashev-Musatov, O., and Shvartzburd, S. (2000). *Algebra i matematicheskii analyz. Uchebnik dlya 10 klassa fiziko-matematicheskogo profilya* [*Algebra and Mathematical Analysis. Textbook for the 10th Grade of Physics and Mathematics Specialization*]. Moscow: Mnemozina.

Vishniakova, S. M. (1999). *Professionalnoe obrazovanie: slovar. Kliuchevye ponyatiya, terminy, aktualnaya leksika* [*Professional Education: Dictionary. Key Concepts, Terms, Contemporary Lexicon*]. Moscow: NMTS SPO.

9

Russian Influence on Mathematics Education in the Socialist Countries

This chapter explores the influence of Russian mathematics education on education in countries within the USSR's sphere of influence — the socialist states. An attempt to describe "socialist mathematics education" has already been made by Swetz (1978). Naturally, no one ever decreed that mathematics education in the countries of the so-called "socialist camp" should be identical. Yet political connections did give rise to connections in culture and education, and consequently the influence of Russian mathematics education has probably been the strongest in precisely these countries.

The discussion below will address three countries: Poland, Hungary, and Cuba. Each of these countries has its own history of relations with Russia and the Soviet Union. Principalities that once were part of Kievan Rus' later entered the Kingdom of Poland and Lithuania, influencing the other parts of this kingdom and being influenced by them in turn. Poland in the 17th and even the 18th centuries was a significant source and channel of learning for Russia. Subsequently, Poland entered into the Russian empire, and lived in many respects the same life as the other parts of that empire. A brief period of complete independence gave way to a period during which Poland was part of the Soviet bloc. Hungary's history of inclusion within the Russian (Soviet) sphere of influence is much shorter: it began only after the Second World War. The history of ties between Russia and Cuba is shorter still, beginning only in the late 1950s. Each of these three countries has its own history of mathematics education, with its own achievements, which are briefly described below. However, readers will also find considerable parallels between what happened in these countries after they came within the Soviet sphere of influence.

In general, the role of foreign influences in education is an important and not fully researched topic. Political ties and antagonisms by no means always translate into preferences in mathematics education (see, for example, the attitude of Italian mathematics educators to the Austrian system of mathematics education (Zuccheri and Zudini, 2007). The three "case studies" provided below facilitate a better understanding of the ways in which national traditions of mathematics education in these three countries interacted with Russian (Soviet) influences, traditions, and materials, which ultimately became — although, of course, only to a limited extent — a part of the national traditions of mathematics education in these countries.

References

Swetz, F. (1978). *Socialist Mathematics Education*. Southampton, PA: Burgundy Press.

Zuccheri, L., and Zudini, V. (2007). Identity and culture in didactic choices made by mathematics teachers of the Trieste section of "Mathesis" from 1918 to 1923. *International Journal for the History of Mathematics Education*, *2*(2), 39–65.

9.1

The Traditions and Development of Mathematics Education. Case of Poland

Antoni Pardała

Department of Mathematics,
The Faculty of Mathematics and Applied Physics,
Rzeszów University of Technology, Poland

1 Introduction

The scientific objective of this section is to answer two questions:

(1) What is Polish tradition in mathematics education?
(2) What is the impact of Russia and the USSR on this tradition and the development of mathematics education in Poland?

The history of Russian–Polish relations is full of dramatic events. It is clear, however, that these two neighboring countries exerted a substantial influence on one another over the course of several centuries in many different spheres. Mathematics education is no exception. It is not possible to grasp the influence of Russian mathematics education in Poland without recognizing the Polish mathematics education tradition itself. The aim of the present article is to arrive at an understanding of this interconnection.

During the second half of the 20th century and especially in recent years, there has been a growing interest in studying the historical roots and developmental processes of mathematics education worldwide, including Poland. A positive outcome of the trend were new scientific publications as well as textbooks on the history of mathematics and the philosophy of mathematics intended for students and teachers of mathematics. Among them is the periodical *Antiquates of Mathematicae* published since 2007, this journal is a continuation of the 19 volumes of materials from the All-Poland Schools on the History of Mathematics, organized by various public institutions in 1987–2006.

The present article is based on the analysis of the existing relevant literature and the author's personal communications with people who were major contributors to Polish mathematics during the second half of the 20th century. The present paper attempts to formulate a synthetic description of the Polish traditions of mathematical thought. This historical perspective guides the analysis of the effects of these traditions on the beginnings of the theory and practice of teaching mathematics and, further, on the development of didactical research in mathematics. At the same time the paper highlights the determinants and difficulties hampering the qualitative development of mathematics didactics in Poland and the blossoming of the Kraków School of Mathematics Didactics under Professor A. Z. Krygowska. The concluding part of the paper provides some reflections and comments on the "ups and downs" of the development of Polish mathematics didactics in the past and in the present.

2 Polish Mathematics Education before the Partition of Poland

The birth of the Polish mathematics education tradition and practice is interwoven with the development of mathematics teaching in the world and also with the historical relations of the Polish state with its neighboring countries. Obviously, credit should be given to the Church for its support of education in the Middle Ages. The Church preserved and accumulated knowledge and traditions of ancient culture after the fall of the Roman Empire. The Church founded schools in the Middle Ages; however, as R. Duda (1983) writes, *interest in mathematics in those schools was practically nonexistent.* The first signs of the weakening of church supremacy in education appeared when universities came into existence throughout Europe. New awareness of and the need for articulating secular aspects in education responded to the general needs of society. This necessitated a change in the previous perception of mathematics and more concern for its content and methods. It also made clear that the modernization of mathematics teaching was inevitable. Successful sea voyages, technological progress, economic recovery, and interchange of goods between nations stimulated educational development, including the development of mathematics education.

The development of mathematics education in Poland was part of the broader process of the development of mathematics education throughout Europe. Until the 17th century, Poland could not boast of any remarkable mathematical achievements (Więsław, 1997). Mathematics teaching in Polish schools in the 16th and 17th centuries was at a basic level and utilitarian in nature. Polish handbooks of arithmetic did appear in that period, however, by authors such as Kłos, Wojewódka, Schedel, and others; a few outstanding mathematicians of the period were recognized such as Jan Brożek, Stanisław Pudłowski, and Adam Kochański.

Education reforms in Prussia, France, and subsequently in other countries — stemming first and foremost from the increasing role of government in education — were echoed in Poland also. A complex teaching reform in Poland was pioneered by the priest Stanisław Konarski, whose ideas were based upon the valuable experience of educational reformers in Prussia and Austria as well as the content of foreign works and handbooks of mathematics. To his great credit, Stanisław Konarski founded the Collegium Nobilium in 1740, which became a school for future state dignitaries. He reformed Piarist schools (followed by a reform of Jesuit colleges). He modified teaching concepts and established principles for creating and implementing the curriculum. Most importantly, he introduced the Polish language into the schools.

Another manifestation of the movement to improve the quality of education was the founding of the Cadets' Knight School (in 1765), where great attention was paid to the selection of valuable handbooks and mathematical literature, maintaining a qualified mathematics teaching staff, and adhering to high standards in the teaching of mathematical subjects. The well-known mathematicians who worked there included Krzysztof Pfleiderer, Józef Łęski, and Michał Jan Hube, who published scientific papers on physics and mathematics. During the time of the reforms, the prestige of the mathematical sciences, applied mathematics, and mathematics education in secondary schools, colleges, and universities grew considerably.

Despite these developments, the situation in the mathematics education of Polish society changed dramatically only when in 1773 after the dissolution of the Society of Jesus, the Parliament (Sejm) established the National Committee of Education (KEN). The Committee is considered to be the first Ministry of Education in the world. The regulations issued by the Committee for provincial schools in 1774 "placed mathematics first among school subjects" (Massalski, 1987). A comprehensive reform of the Polish education system at all levels (including Kraków University and Vilna Academy) was proposed that included:

- the introduction of the Polish language as the language of instruction in schools,
- the reorganization of secondary schools and an extension of the time of education from four to seven years,
- changes in teaching methods and curricula,
- preparation of new textbooks in arithmetic, geometry, and algebra, selected through an open competition, and
- training teachers for national schools.

The reforms initiated by the National Committee of Education were implemented in stages. In 1783, KEN issued regulations related to the reorganization and reform of academic education. The aim was a comprehensive reform *prescribed for the academic profession and schools in all regions of the Republic of Poland*. A significant achievement of KEN was the creation of the Society for Elementary Books as an institution responsible for supplying national schools with textbooks. The competition for writing textbooks in arithmetic, geometry, and algebra for national schools was won by Simon Lhuillier, a Swiss mathematician. His texts were translated into Polish by Andrzej

Gawroński. Other arithmetic, algebra, and geometry textbooks by such authors as A. Dąbrowski and I. Przybylski also were approved by KEN and then used in school practice.

The reforms introduced by KEN "had a greater chance of success than the educational reforms previously introduced in both Poland and other countries," writes Z. Pawlikowska-Brożek (1982), but their implementation was interrupted by the successive partitions of Poland in 1793 and 1795, when the independent Polish state ceased to exist and parts of the country fell under Russian, Austrian, and Prussian rule.

3 Polish Mathematics Education in the Part of Poland under Russian Rule

Using the history of secondary schools in the town of Kielce as an example, A. Massalski (1987) analyzed the mathematics teaching experience and practice in the part of Poland under Russian rule during the late 18th and 19th centuries. Massalski (1987) explored how schools in Kielce came into being, how they functioned, how they were closed and re-established. For many years, there was only one secondary school in Kielce, which was run by priests, and it was only in 1816 (or even few years later) that another secondary school, the Departmental School, was established (its name was changed to the "Provincial School" in 1819). This school was supported and managed by the state.

Education in such schools lasted seven years and ended with matriculation. After the uprising of 1830 was crushed, repression and Russification of secondary education in the Kingdom of Poland increased. The employment of Russians as secondary school teachers was strongly encouraged, and the right to teach in state schools, in practice, was given only to those Poles who had graduated from Russian universities.

Further changes in the functioning of the secondary schools in the Kingdom of Poland were brought about by the *Wielopolski Law of 1862*, which transformed schools of a vocational nature into classical gymnasiums with seven-year programs. Their curriculum included 37 hours of mathematics per week. This law was passed during a period of liberalization in the Russian empire and, not surprisingly, its educational curriculum was strongly rooted in Polish culture. After the Defeat of the January Uprising (1863), the Wielopolski Law was partly altered in 1864 and then totally annulled by Russia in 1866. It was replaced by new education regulations. These administrative efforts laid the groundwork for the full Russification of the schools, which included the introduction of the Russian language and Russian textbooks into mathematics classes.

In 1873–1890, secondary school education in the Kingdom of Poland was carried on in accordance with the regulations of 1871 developed by the Russian Minister D. Tolstoy. These same regulations governed education across the Russian empire. The Soviet historian Ganelin (1954, p. 69) wrote about this period: "In general, the reform of 1871 should be considered a great evil in the history of the Russian school, one of its darkest and most disgraceful episodes." Polish education, just as education in other parts of the Russian empire, suffered from features of the new educational system such as overloading the curriculum and the excessive and petty micromanagement of all aspects of school life.

4 The Period of the International Reform Movement and the Independence of Poland

In the beginning of the 20th century, the Meran program led by Felix Klein was a source of inspiration for mathematics education reform in European countries. Even though Poland was still under foreign rule, during that period, there was a revival of activities of Polish educators, mathematicians, and mathematics teachers aimed at a modernization of mathematics teaching methods. The Mathematics and Physics Circle — with its first president, S. Dickstein — as well as the journal *Wektor* and other journals for mathematics teachers became widely known. Attempts to improve mathematics education resulted in the creation of a working document in 1919, which was entitled "The Educational Programme for Secondary Schools." This document was used as a departmental proposal for how to organize secondary schools in Poland after regaining independence and as a basis for mathematics teaching programs that were published in 1919–1922.

The achievements of Polish pedagogical thought in mathematics during the first half of the 20th century are carefully analyzed by W. Dubiel (1996). The author focuses on two categories of pedagogical thought, the "theoretical" and the "intuitively practical," as well as on two key questions posed in discussions of mathematics education: (1) what kind of mathematics should be taught? and (2) to what extent should school mathematics be rigorously scientific? During the rebirth of Polish statehood, a number of valuable books on mathematics didactics were written by O. Nikodym, S. Neapolitański, W. Nikliborc, and others. The movement to modernize mathematics education and popularize mathematics didactics was backed by creative teachers (S. Steckel, I. Zydler, and others) and outstanding mathematicians such as S. Banach, W. Sierpiński, H. Steinhaus, S. Straszewicz, S. Saks, A. Tarski, W. Wilkosz, A. Łomnicki, and others. They were the authors of academic or secondary school mathematics textbooks and books on mathematics teaching for the general public. They also published articles in periodicals for teachers, presented reports at Meetings of Polish Mathematicians, and gave lectures at methodology courses organized for mathematics teachers.

In that period a valuable contribution to the development of world mathematics was made by Polish mathematicians and Polish mathematical schools in Warsaw and Lvov (Wachułka and Dianni, 1963). Progress in Polish mathematical work was interrupted by the Nazi occupation, which reduced mathematics teaching to clandestine classes. After the Second World War, Poland found itself in the USSR's area of influence.

5 Mathematics Education in Poland after World War II and before the Collapse of the Soviet Union

The political, economic, and social changes that took place in Poland after World War II gave rise to new approaches to mathematics education in theory and practice. The country had to prepare engineers, teachers, and experts in new fields in order to recoup the losses brought about by the war. This posed new problems for mathematics education. Importantly, after the elections to the Legislative Sejm in January 1947, the entire educational process became deeply ideological. That was followed by the successive introduction of new standards for organizing the educational system and institutions of higher education. A new kind of education of mathematics teachers also was developed.

The prewar standards for the organization of the school system, curricula, school-level mathematics textbooks, etc., were in effect until a transitional program of mathematics education was published in 1949: *The Teaching Programme for 11-Year Secondary Schools (Draft). Mathematics, Warsaw 1949.* In the 1948–1949 school year, a new, 11-year school was introduced; it included the four-year secondary school and the seven-year primary school.

At the same time, learning Russian became obligatory in universities and in schools, beginning in the fifth grade. Learning other foreign languages was non-compulsory. At the same time private schools were closed and the teaching of religion was gradually eliminated from schools. Many educational institutions were closed or taken over by the state. The ideologization of education — which went against Polish tradition — continued to grow. Education in schools and colleges was to be based on Marxism–Leninism. The school and the college were expected to inculcate in students a firm conviction about the superiority of socialism and its overwhelming advantages, to bind patriotism and internationalism, and, in particular, to encourage students to accept the Soviet Union as the main partner and ally of the Polish People's Republic (PPR). Large-scale reliance on Russian aid and an "ascent" to Soviet standards followed. In particular, mathematics education literature from the Soviet Union was disseminated broadly in Poland and was highly influential. At present great pride is taken in various achievements which resulted from the school and curriculum reforms carried out then. There is no regret about the idea of the introduction of the 10-year secondary school (Soviet standard) heralded by the 1973 Sejm resolution.

Historically, creative cooperation between Russian and Polish mathematics schools, in particular, represented by W. Sierpiński and

N. N. Luzin and their disciples was quite good. At this time the collaboration in research mathematics developed even further. It was accompanied by collaboration in the field of mathematics education. Polish mathematicians and mathematics educators were educated in their specialties in Soviet universities and academic institutions — where they obtained scientific training and conducted doctoral and postdoctoral studies — supported by the state. Mathematical and mathematics education literature was available in Russian in Polish bookstores that specialized in international books and periodicals, and could be acquired relatively cheaply by Polish readers, including students, teachers, and academic staff. The privilege of being selected to study at Soviet mathematical institutions, offered to Polish students and young scholars, was considered very attractive. Intensive cultural cooperation in secondary school and university circles was facilitated by organization of inexpensive study tours and camps in the USSR.

During these years, an active school of methodology developed in Poland. This school supported the restructuring of mathematics education in the Polish People's Republic. Its principal members included T. Gutowski, B. Iwaszkiewicz, S. Kartasiński, A. Z. Krygowska, S. Kulczycki, A. Rusiecki, and S. Straszewicz. The activity of this school was closely linked both with the work of Western mathematics educators and their organizations (in particular, with the Commission for the Study and Improvement of Mathematics Teaching — CIEAEM). Polish mathematics educators shared a growing belief that scholars and professionally active teachers needed to begin cooperating and introducing changes in mathematics teaching and the education of mathematics teachers. Such initiatives required the involvement of creative mathematicians and the support of authorities and decision-makers. An important role in this respect was played by the Polish Mathematical Society as well as the activities of professor A. Z. Krygowska and her Kraków School of Mathematics Didactics (Nowecki, 1984; Ciosek, 2008). It appears that Poland, in its turn, functioned as a kind of channel through which certain new methodological ideas penetrated into other socialist countries, including the Soviet Union.

An important role in the dissemination of these new ideas was played by the journal *Dydaktyka Matematyki* (renamed *Didactica Mathematicae* in 2007). Its first volume was published in 1982 and its editor-in-chief was Professor Anna Zofia Krygowska. The journal was unique for countries within the Soviet sphere of influence and turned out to be very important for the international dissemination and popularization of the ideas and achievements of the emerging scientific circles of mathematics didactics in Poland and of the Kraków School of Mathematics Didactics in particular.

6 Conclusion

Polish mathematics education had its inception and initial development within the context of a broader European tradition. At a certain stage, a large part of Poland became part of the Russian empire and consequently mathematics education began to be governed by the same laws, and to employ the same textbooks and problem books, as the rest of that empire. It may be argued, however, that the coercive Russification of that period, for the most part, achieved the opposite of its desired effect, encouraging Poles to take special care to preserve Poland's distinctive character. In turn, Poland also influenced the Russian education sometimes. In particular, in 1803 the Russian empire reformed its educational system, including universities, using as exemplars the Polish reforms. This reform was carried on mainly by Poles (Więsław, 2007).

Russian (Soviet) influence during the years following World War II had a different effect. It was exerted through textbooks and other books, through continuing education courses for teachers, etc., relying on organizational structures that were largely modeled on Soviet prototypes. Polish schools of methodology, however, continued to rise and function in Poland, and these schools not only took into account what came from the USSR, but also remained aware of what was happening in the West as well as in Poland's own traditions. Openness to different methodological ideas has been extremely beneficial for Polish mathematics education.

References

Ciosek, M. (2005). O wpływie Anny Zofii Krygowskiej na dydaktykę matematyki w świecie. *Dydaktyka Matematyki*, *28*, 129–154.

Dubiel, W. (1996). Szkic rozwoju polskiej myśli dydaktycznej matematyki w I połowie XX wieku. *Wyd. Uniwersytetu M. Curie-Skłodowskiej*, Lublin, 1–96.

Duda, R. (1983). O programie szkolnym z matematyki trochę historycznie. *Oświata i Wychowanie Wersja B*, *18*, 49–51.

Domoradzki, S. (1999). Matematyka w Uniwersytecie Carskim w Warszawie w latach 1869–1915, *XII Szkoła Historii Matematyki*. Kraków, 139–151.

Ganelin, Sh, I. (1954). *Ocherki po istorii sredney schkoly v Rossii [Notes on the History of Secondary Schools in Russia]*. Moscow: Uchpedgiz.

Massalski, A. (1987). Nauczanie matematyki w szkołach średnich Kielc w XIX wieku. *Wiadomości Matematyczne*, *27*(2), 293–311.

Nowecki, B. (1984). *Krakówska szkoła dydaktyki matematyki*. Kraków: WN WSP, 1–54.

Pawlikowska-Brożek, Z. (1982). Matematyka w okresie reform szkolnych sprzed lat dwustu. *Wiadomości Matematy*czne, *24*(1), 95–105.

Wachułka, A., and Dianni, J. (1963). *Tysiąc lat polskiej myśli matematycznej*. Warszawa: PZWS.

Więsław, W. (1997). *Matematyka i jej historia*. Wyd. Opole: NOWIK.

Więsław, W. (2007). *Matematyka polska epoki Oświecenia*. Warszawa: Fraszka Edukacyjna.

9.2
Case of Hungary

Katalin Fried

Faculty of Science, Institute of Mathematics,
Eötvös Loránd University, Hungary

1 Introduction

As in other countries, the history of mathematics education in Hungary is embedded in the general history of the country. For this reason a brief historical review of Hungary and its national educational system is necessary. The Hungarian people have seen many dramatic changes (Molnár, 2001). In the 15th century, the once-strong Hungarian nation was conquered by the Ottoman Empire. Following a liberation itinerary, Hungary became a part of the Austro–Hungarian Monarchy. After the unsuccessful revolution of 1848, Hungary remained within the empire; however, the compromise of 1876 granted the Hungarian people more autonomy within it. This period, from 1876 until 1914, often is considered to be the golden age of the Hungarian nation. After World War I, Hungary became fully independent, but suffered from many political and economic problems. In this situation following the German policy seemed attractive to many Hungarian leaders.

As a result of World War II, Hungary was occupied by the Soviet army. In a period of a few years, much of the population was forced to change social status. Institutions and entire sectors of the society were eliminated (Szendrei, 2007). The revolution of 1956 against the Socialist government supported by the Soviet army, although ultimately unsuccessful, helped establish a regime which was more liberal than in other Soviet bloc countries. It was only in 1989, however, that the Soviet army withdrew from Hungarian soil.

In the second half of the 19th century, education became one of the focal points of Hungarian policy (Nagy, 1993). Both the government and Hungarian society in general were very concerned with educational development. This widespread interest resulted in the creation of several important innovations which were supported under the later government. Particular attention was paid to science and mathematics education. Baron József Eötvös who served as the Hungarian Minister of Education became famous and remains so today because of his support for the law that established the right (and, indeed, the duty) of all citizens to attain a minimum level of education in reading, writing, and computational skills. His son, Loránd Eötvös, a physicist whose Ph.D. advisor was Kirkhoff, served as the Minister of Education for a brief period and founded the Eötvös-Kollégium. This was modeled on the French École Normale Superieure. He named the institution after his father. The purpose of this Kollégium was to increase the number of well-qualified teachers of science. It proved to be very effective (Vogeli, 1997). The importance of the field of education was acknowledged by the inclusion, in 1870, of professional studies in education within the Hungarian university curricula.

2 Mathematics Education before WWII

The tradition of gifted education in Hungary originated in the late 19th century. This Hungarian model proved to be of importance in other countries, including the Soviet Union. The most famous professors of the Budapest Technical University — Gyula König, Gusztáv Rados, József Kürschák, Mór Réthy, and others — recognized that it was extremely important to improve mathematics education and particularly, to organize and develop a specialized approach to teaching talented students. A committee was established to design a better curriculum in mathematics. Gyula König and Mór Kármán were important participants. In 1906, under the leadership of Manó Beke, a new Hungarian mathematics curriculum was adopted.

Another major step was made by Dániel Arany, founder of the *Mathematical Journal for Secondary School Students (KöMaL)*. The journal not only published problems, but also organized contests in which the students sent solutions to the Journal. For every problem the editors published the best submissions.

Most importantly, the journal published the names of the students who submitted the correct solution. This proved an important way to motivate the brightest students. Additionally, the journal published articles on different mathematical topics that widened the interests of the students. Many famous mathematicians and teachers of mathematics served as editors-in-chief of the journals including Dániel Arany, László Rátz, János Surányi, Ervinné Fried, and others. This journal has continued to the present. Almost all famous Hungarian mathematicians started their career as problem solvers of KöMaL.

Another important tradition also emerged at the end of the 19th century — the tradition of mathematical competitions. These traditions were later supported and developed in many countries, including the Soviet Union and the United States. The problems posed in these Hungarian Olympiads are still an important source for scholars interested in mathematical competitions and gifted education (Kürschák, 1963; Hajós *et al.*, 2001). Due to the high level of Hungarian mathematics education before WWII, Hungary emerged as the world's leading contributor of research mathematicians in proportion to its population.

3 Hungarian Mathematics Education after WWII

As a result of the war the Hungarian people suffered many losses and a severe shortage of qualified teachers. Teachers of mathematics were in particularly short supply. One of the reasons for this was that individuals with mathematical training were needed elsewhere to support the reconstruction of the country; another was that the number of lessons in the humanities was cut, while the number of lessons in mathematics was increased. The government encouraged teachers of other subjects to teach mathematics and also provided a "fast track" education for mathematics teachers. These teachers generally were less qualified than those with better preparation, although some of them performed well.

Among other post-war shortages was that of textbooks and collections of problems. Indeed, pre-war textbooks had been in use in schools for a very long time and often were in very bad conditions. Importantly, KöMaL, which had been a source of high-level problems before the War, was not published during the war and did not resume publication until 1947.

Beginning in 1946 fundamental changes in the Hungarian school system were under way. Especially important was the creation of an eight-year "general school" providing equal opportunities for every student. Halmos and Varga (1978) reported that the number of teachers increased from 30,000 in 1937/1938 to 66,000 in 1960/1961. They mentioned that the new system provided much less knowledge to *every* student than the previous system had provided to the *highly selected* population. The difference in numbers of those receiving this eight-year education was significant however.

4 Russian Textbooks and Problem Collections

Mathematics teachers tried to use all available books to overcome the limited resources for practicing problem solving. In the meantime, more and more Russian texts and problem books became available. It is important to mention that Russia was a significant channel for receiving literature from other countries as well — often even non-Russian books were translated into Russian and then into Hungarian, because translation from Russian could be arranged more easily than translation from their original language.

Probably, the most popular collection of algebraic problems for high schools was by Larichev (1952). Despite its widespread use, "Larichev" is currently criticized by many Hungarian teachers who taught using the book (personal communications). According to its critics, the book was focused upon students' mastery of calculations and algebraic transformations. Less attention was paid to the development of thinking. Another issue was that the word problems in the book glorified peasants and workers, factories and socialist production, etc., which was viewed by Hungarian teachers as "Soviet propaganda." The book, however, was in use from 1952 until the late 1970s and definitely influenced the process of education. Its importance was enhanced by the reputation of its translator, Tamás Varga, one of the most important Hungarian mathematics educators of the post-war period.

The use of Russian mathematics literature was not limited to just this book. Russian books for higher education in mathematics and for gifted students were used widely (for example, Shkliarskii, Chentsov, and Iaglom, 1970). Later, when the Russian journal *Kvant* was founded for students interested in mathematics, it became a source for some articles and problems that appeared in KöMaL.

Important Hungarian mathematics educators of that period, including Imre Rábai, Lóránt Pálmai, and Endre Hódi, were influenced not only by the content of some Russian books, but also by their methodology (personal communications). Some of these books were considered to be exemplary in constructing clusters of problems leading to discovery of important connections between different topics in mathematics, or in enabling students to understand sophisticated situations and statements. These books were designed to build up a topic in mathematics in such a way, that by solving consecutive problems, the readers themselves could construct a theoretical understanding of the topic.

While many Soviet textbooks exerted a positive influence on mathematics instruction in postwar Hungary, the system of teaching became much more rigid than it had been before WWII. The teachers had little freedom of choice. The curriculum was strict, and the teachers had to teach their lessons in conformance with the official syllabus and schedule. Teaching was strictly monitored and controlled. This "centralization" was also viewed as Soviet inspired, although the role of Hungarian authorities obviously was not insignificant.

5 Reform of Postwar Curriculum

The need for changes in the curriculum was recognized in Hungary very early. It was in 1949 that well-known mathematicians Rózsa Péter and Tibor Gallai (1949, 1950) started publishing their textbooks for high school students. Their books were based on modern mathematical principles and approaches, and were predecessors in some sense of the

"New Math" texts in the United States. Russian textbooks seem not to have influenced these Hungarian inspired textbooks. On the contrary, the Hungarian tradition of textbook writing can be observed clearly in these texts. The books had a limited success, however, because they proved to be too challenging for both students and teachers.

Hungarian textbooks written later and, in particular, those connected with the reforms lead by Varga, also were rooted in Hungarian tradition, but seem to be more connected with the Russian experience. Varga (1988) said that his sources of inspiration were Dienes and Krygovskaya (rather than Russian educators). It is important, however, that mathematics education reform in Hungary started later than in the West. "Boom of the new math with its fresh wind and dust storms did not reach us before some years later," wrote Halmos and Varga (1978, p. 225). Chronologically, Hungarian reforms followed reforms in the Soviet Union initiated by Kolmogorov.

It is known that there were regular contacts between Soviet and Hungarian mathematicians and mathematics educators in general and between Kolmogorov and Varga in particular. For example, Hungarian mathematician M. Arató (2006) reports that Kolmogorov familiarized himself with some Hungarian reform ideas while in Hungary. Arató, the applied mathematician, reports that Kolmogorov found Hungarian mathematics education even more abstract than that in the Soviet Union which Kolmogorov promoted and supported. According to Arató, Kolmogorov recommended that teaching students counting should not be forgotten, and, also, that more attention should be paid to general education rather than to the education of the gifted.

The important difference between Russian and Hungarian reform was that the Hungarian mathematics reformer concentrated on elementary and middle school rather than on high school.

6 Classes for Talented Students

Hungary had a well-developed tradition of gifted education, particularly in gymnasiums formerly affiliated with the Eötvös-Kollégium. An important step in its development was connected with the Russian (Soviet) experience. The idea of creating classes for talented students in

Socialist Hungary was discussed for the first time after the emergence of this idea in the Soviet Union.[1] Imre Rábai and his colleagues, who wanted to initiate the beginning of such a class in Hungary, started discussions of that possibility in the Ministry of Education in the late 1950s. This suggestion was rejected at this time. Later, in 1962 for some reasons still unknown to Rábai (2009) the decision was changed and he was invited to organize and teach such a class. Interestingly, he was allowed to organize the class not in the school which he suggested originally, but in another one, the now most famous gymnasium "Fazekas."

The decision was made in August so Rábai had hardly enough time to pick 30 students for the class. Luckily, he was able to invite to the class the winners of mathematical competitions. That class can be considered to be the best mathematical class in Hungary. The class included Lovász, Laczkovich, Pelikán, Bollobás, and Pósa, to mention just a few. Even those who did not become mathematicians became well-known in other fields of science.

After the success of the first special mathematics class, other schools received permission to start special classes also (for example, Berzsenyi and István gymnasiums in Budapest). Russian texts and problem books and the Russian experience no doubt were very important for Hungarian special classes for the mathematically talented.

7 Conclusion

The influence of Russian mathematics on the teaching of mathematics in Hungary was twofold. First, direct methodological influence was indeed significant. Despite rich Hungarian traditions of mathematical education, ideas and materials coming from Russia were of importance. Given that the traditional pre-war system was reorganized so that

[1] *Editorial note:* The influence of pre-war Hungarian schools on Soviet educators was in turn also very substantial (although these Hungarian schools were not officially called schools for the mathematically talented). One of the editors of this book remembers Kolmogorov publicly attributing the origin of the idea of special schools for the mathematically gifted to Hungary (Vogeli, 1997).

previously influential social groups found themselves in a very different situation, the Russian experience of universal but still challenging education was acceptable to the Hungarian people. The exchange between Russia and Hungary of educational materials, methodologies, problems, etc. was substantial.

Both in Russia and Hungary, the role of mathematicians in devel oping and establishing standards of mathematics education was very important. The influence of mathematicians was particularly visible in the education of the mathematically gifted. Additionally, Hungarian mathematicians trained in Russia consciously or unconsciously brought some Russian educational ideas back to Hungary.

The second kind of influence, more political than methodological, cannot be overlooked. Hungarian authorities always had to verify that their policy was aligned with that of its "senior brother." That does not mean that cultural or educational decisions were always identical to a Soviet prototype. In some situations, Hungarian policy was even the opposite of the Soviet. Still, it can be argued that the creation of Russian schools for the mathematically gifted or the reform of the mathematical curriculum in the Soviet Union provided a few additional "green lights" for similar initiatives in Hungary.

The political situation changed dramatically after 1989. Hungary became much more open to the world, including Western countries. Many previously important issues, lost their value. Still, the connection with Russian mathematics education easily can be identified. Moreover, sometimes it is not possible to distinguish between original Hungarian materials and problems and those which came from Russia and now have become a part of the folklore of Hungarian mathematics education.

References

Arató, M. (2006). A. N. Kolmogorov v Vengrii. In A. N. Shiryaev (Ed.), *Kolmogorov v vospominaniyakh uchenikov* (pp. 28–33). Moscow: MTSNMO.

Fried, E. (2009). Personal communication with the author.

Hódi, E. (2009). Personal communication with the author.

Hajós, G., Neukomm, G., Surányi, J., and Liu, A. (2001). *The Hungarian problem book. Based on the Eötvös competitions 1929–1943.* Washington, DC: Mathematical Association of America.

Halmos, M., and Varga, T. (1978). Change in mathematics education since the late 1950's — ideas and realisation. *Educational Studies in Mathematics*, *9*, 225–244.
Kürschák, J. (1963). *The Hungarian problem book. Based on the Eötvös competitions 1894–1905 and 1906–1928* (2 vols.). New York: Random House.
Larichev, P. A. (1952). *Algebrai feladatok gyûjteménye*. Budapest: Közoktatásügyi Kiadóvállalat.
Molnár, M. (2001). *A Concise History of Hungary* (Cambridge Concise Histories). Cambridge-New York: Cambridge University Press.
Nagy, P. T. (1993). *The Outlines of Hungarian Education 1500–1945*. Budapest: Educatio Pub.
Pálfalvi, J. (2009). Personal communication with the author.
Pálmay, L. (2009). Personal communication with the author.
Péter, R., and Gallai, T. (1949, 1950). *Matematika a gimnázium I. és II. Osztálya számára*. Budapest.
Rábai, I. (2009). Personal communication with the author.
Shkliarskii, D., Chentsov, N., and Yaglom, I. (1970). *Geometricheskie neravenstva i zadachi na maksimum i minimum*. Moscow: Nauka.
Szendrei, J. (2007). When the going gets tough, the tough gets going. Problem solving in Hungary, 1970–2007: Research and theory, practice and politics. *ZDM Mathematics Education*, *39*, 443–458.
Vogeli, B. R. (1997). *Special Secondary Schools for the Mathematically and Scientifically Talented. An International Panorama*. Teachers College, Columbia University.

9.3
Case of Cuba

Orlando B. Alonso

Teachers College, Columbia University, NY, USA

1 Introduction

The year 1959 marked the beginning of a new period in the history of Cuba characterized by radical changes in all areas of Cuban culture. The influence of the communist doctrine of social revolutions as the unique solution to national social problems in a politically polarized world took roots in Cuban soil. After 1959, the Soviet Union's influence in economic policy, cultural life, and education changed the life of Cuban people.

This paper is an account of Russian influence on mathematics education in Cuba after 1959. It begins with a summary of Cuban traditions in mathematics education prior to the 1959 Revolution, followed by an examination of changes in the organization and pedagogy of mathematics education in secondary schools, higher education, and teacher training.

2 Cuban Mathematics Education before 1959

Mathematics education programs in Cuba before 1959 existed in either the public education system or in private schools and colleges often founded by Catholic religious orders, or institutions that had evolved from a colonial schooling tradition. The state maintained free public education at all levels. Pedagogical degrees were awarded in special schools for elementary teachers and schools of pedagogy in the universities.

At the first National Congress of Mathematics celebrated in Cuba in 1982, Dr. Carlos Sánchez Fernández and Dr. Concepción Valdés Castro presented a historical sketch of the development of the mathematical culture in Cuba (Sánchez and Valdés, 2003), emphasizing the contribution of the most prominent personalities involved in this process. Their accounts of the historical period between 1900 and 1959 are summarized in this section.

According to Sánchez and Valdés, the year 1900 marked the creation of the first School of Pedagogy in Cuba at the University of Havana following the Spanish–American War. Cuba's most prominent pedagogue of the period, Dr. Enrique José Varona (1842–1933), proposed the creation of the school. The University of Havana, which was founded in 1728, maintained two chairs, one in Mathematical Analysis and the second in Geometry and Trigonometry. At the time of the reform Dr. Claudio Mimó y Caba (1844–1929), who was born in Cataluña, Spain, was appointed Titular Professor. Dr. Mimó y Caba was the founder and director of a prestigious secondary school. At the University of Havana, Dr. Mimó y Caba was both the teacher and sponsor of the first Doctor in Physics–Mathematics Sciences candidate after 1900, Pablo J. Miquel y Merino (1887–1944). Miquel was the most influential figure in Cuban mathematics education during the first half of the 20th century. He was especially influential at the university level.

Dr. Miquel was born in Havana and attended the Jesuit College, the most prestigious institution in the Cuban educational system. In addition to his doctoral degree obtained in 1908, Dr. Miquel also graduated from the University of Havana as a Civil Engineer in 1910 and as an Architect in 1912. In 1913, he was appointed the head of the Analysis Mathematics chair at the University, where he remained until his death 30 years later. Dr. Miquel's books in mathematics teaching at a university level were considered among his major contributions. The *Elementos de Álgebra Superior* (1st Edition published in 1914, 2nd Edition published in 1939, and 3rd edition published in 1943, with 695 pages), and his book *Curso de Cálculo Diferencial e Integral* (Volume 1 appeared in 1941 and Volume 2 in 1942) were widely used in universities and for the introduction of new courses and textbooks supporting mathematics changes that occurred in the program of studies at the end of the 1930s and the beginning of the 1940s. In 1942, Dr. Miquel was among the founders of the Cuban Society of Physics and Mathematics Sciences and was its first president. He remained president until his death.

Following the creation of East University in 1947 and Central University in 1952, the mathematical and scientific level of the country increased rapidly. By the late 1950s, 120 new doctorates in the physical–mathematical sciences had been added to the list of 36 doctorates who were graduated during the first 33 years of the republic. This increased professional manpower and solidified an emergent national tradition in mathematics education.

Among the new generation of Cuban mathematicians, the most outstanding figure with wide international influence was Dr. Mario Octavio González Rodríguez (1913–1999). Born in Matanzas, he was a secondary school student of the eminent mathematician and pedagogue, Dr. Manuel Labra Fernández (1900–1982). Dr González was appointed Auxiliary Professor of the Department of Mathematical Analysis of Havana University in 1940, succeeding Dr. Miquel as chair and Titular Professor. He wrote *Algebra Elemental Moderna,* published by Editorial Selecta, Habana in 1956–1957 in two volumes, and also *Complementos de Aritmética y Algebra* and *Complementos de Geometría y Nociones de Cálculo Diferencial e Integral.* The 9th and 10th edition, published in the 1960s, played a fundamental role in the growth of mathematical culture in Cuba before and after 1959.

One of the most influential Cuban mathematics educators during the last two decades was Dr. Aurelio Baldor (1906–1978), pedagogue, lawyer, and founder of a prestigious private college. In 1941, he wrote a book entitled *Algebra*, known by students as "Algebra Baldor," that was adopted as the official textbook in public and private schools in Cuba until 1961. Arguably the most influential book in mathematics education in Latin America, Algebra Baldor still is reprinted every year in North America, Latin America, and Spain. *Geometría Plana y Del Espacio y Trigonometría*, and *Aritmética* are among other important books written by Baldor.

In addition to native and Spanish born mathematics educators, Cuba profited from assistance from foreign sources even before the collaboration with the Soviet Union following the Castro regime. Dr. Bernard Gundlach, a holder of a PhD from the University of Hamburg, arrived in Havana as a passenger aboard a ship of Jewish refugees from Nazi Germany that had been denied landing rights in the United States. Gundlach had left Germany with his Jewish wife and traveled through France, into Spain, and then to Havana. His background in the history of science and mathematics placed him among the early historians of mathematics education. He learned Spanish quickly and taught mathematics to both children and university students. Both Castro brothers were among his students. Gundlach stressed the importance of the recapitulation process in teaching mathematics, adapting mathematical methods from the history of science for classroom use. Following the socialist revolution, Gundlach left Cuba for America, where he worked as a faculty member at Bowling Green State University.

3 Changes in Organization and Pedagogy after 1959

After the fall of the Batista government and the rise of Fidel Castro, religious and private schooling were immediately dismantled and still are prohibited in Cuba to this day. Private and Church school buildings were seized. Hundreds of teachers were put out of classrooms because of religious beliefs or political ideas considered incompatible with the ideology of the revolution. The numbers of mathematics professors at the University of Havana that remained active were reduced to one. A Cuban born after 1959 may never be aware of the existence of preceding traditions. One educational system was simply substituted for another almost overnight.

Educational dismantlement occurred in a climate of confrontation. The motto "Universities are for Revolutionaries" left no room for dissent in the classroom. Student organization participation in universities in the 1960s led to occasional excesses (Guadarrama, 2005). Ideology penetrated everyday school life. In particular, young communist organizations were created in high schools and universities, while in elementary schools, all students became members of the pioneers' organization, incorporating daily school routines and ideological indoctrination.

The guiding principle of the Socialist Republic was the creation of the "new man." Features generally attributed to the socialist educational system were: the combination of study and work, intellectual preparation and scientific formation, universal access, rewards and depravations determined by political profile, centralization of decision-making processes, and official popular organizations promoting school community and relations (Cruz-Taura, 2003). A pedagogical theory of combining study with physical work attributed to the Russian pedagogue and writer Anton Semyonovich Makarenko (1888–1939), was the basis for opening self-supporting (at least partly self-supporting) residential schools. The Lenin School near Havana is a prominent example. Part of its mathematics curriculum is devoted to optimization of school resources and their management.

Also, in the mid-1960s a system of military schools for thousands of secondary level students was created. Major construction of these schools occurred in the 1970s. Students from these schools continued university education at eight military academies, created in 1977 (Guadarrama, 2005).

The increase of enrollment in the elementary school in the 1960s led to an explosion of enrollment at the secondary school level in the 1970s. Hundreds of secondary school buildings called "Escuelas en el Campo" ("Schools in the Fields") where the principle of the combination of study–work was put into practice, were constructed in every corner of the island. In response to the increased need for secondary school teachers the "Manual Ascunce Domenech contingent" was created. Over five consecutive years, 10th grade students across the nation, after satisfactory completion of 10th grade, were enrolled in a teaching program called "Plan de Formación de Profesores de la Enseñanza General Media" (PFPEGM). Each student worked at a school teaching one class and participating in all departmental and school activities, in addition to taking classes towards a teaching certification. This program extended for five years, after which students were certified as teachers. Teachers who graduated from this program needed only to complete two more years of studies to obtain the equivalent of a bachelor degree in education (Licentiate in Education) in selected fields.

Following the Soviet model, Faculties of Education of the three National universities became Pedagogical Institutes in the mid-1960s and were extended in successive years to the remainder of the provinces, to become independent university centers. One difference between their programs and previous programs derived from the Soviet model is that teachers obtained certification in a single content area instead of dual certification. By the end of the 1970s pre-university "Schools in the Fields" were also fully functional and new buildings were constructed for large vocational schools and for Pedagogical Institutes of Higher Education in all the capital cities of the six provinces according to the old political and administrative system. With the creation of a new political and administrative system, similar institutions were created in the 12 new provinces and the special municipal district of Isla de Pinos. Similarly, other schools that originally were part of universities were redefined as new universities, technical and techno-military institutes (Guadarrama, 2005).

4 Russian Influence in Secondary School Mathematics Education after 1959

Like the school organization itself, Cuban mathematics education was modeled after the Russian (Soviet) system. The curriculum was determined at the national level by ministry documents (MINED, 1982). Control and monitoring became of major importance at each level (national, provincial, and municipal) of education. Special teams of methodologists with the authority to transmit and supervise methodological and curricular implementation were formed in each province.

In addition to their regular classroom teaching, school mathematics teachers were expected to dedicate a number of hours each week to "methodological session" meetings. In these meetings the lesson plans for the coming week were discussed as well as the major methodological aspects of teaching the topics assigned for the next week. These sessions were conducted by the head of the mathematics department who, in turn, received guidance from a municipal methodologist. An important piece of documentation called the *calendar plan of teaching* provided a timeline for teaching each topic and even each lesson. In practice, teachers were strongly encouraged to follow the same calendar plan nationwide as was the practice in the Soviet Union and other socialist countries.

The recommended lesson format also followed the Soviet model. Typically, the lesson was expected to include five stages: introduction, motivation, development, assessments, and conclusions (intended to provide a transition to the next topic). Identical recommendation can be found in Soviet textbooks for mathematics teacher education (Kolyagin *et al.*, 1975).

The curricula itself was also strongly influenced by the Soviet model. The transformation of the school's program occurred with the help of many visiting specialists from socialist countries. For example, Galina Maslova, director of the Mathematics Department in one of the Institutes of Russian Academy of Pedagogical Science, was an influential visitor. These visitors were involved in methodological recommendations and textbook development (see Jungk, 1985, for example). These textbooks as well as textbooks which were prepared later aimed at developing good algebraic and computational skills as well as problem solving abilities. The concept of "developmental education" which originated in works of Soviet psychologists such as Vygotsky became equally important to Cuban educators.

In particular, Soviet influence was very visible in the education of the mathematically talented. Cuban schools for the mathematically talented, including the Lenin School in Havana, emerged following the Soviet pattern, and their curriculum and pedagogical approach were modeled after the Soviet paradigm (Vogeli, 1997). Russian books and problem collections were of importance for teaching in these schools and for all students interested in mathematics. Books from the Soviet Popular Science series written by Soviet academics for high-school students interested in mathematics were especially valuable. These books were translated in Spanish and were very popular among both high school and undergraduate students. Additionally, Russian teachers were imported to assist Cuban students in preparation for participation in Olympiads and Latin American mathematics contests.

5 Russian Influence in Higher Mathematics Education and in Mathematics Teacher Education

The Russian influence can be observed in Cuban higher mathematics education (including teacher education) in many ways. First, the everyday life and system of education in higher education were modeled after that of the Soviet Union. Two formats of teaching were in use: (1) "Conferences" and (2) "Practical Lessons," often taught by different professors. A "Conference" was given by the professor with a higher professional rank who taught in a lecture format sometimes for a class of about 40 students. Later the conference group was divided into two smaller classes for "Practical Lessons," where independent individual work as well as all-class activities were employed. The work in Practical Lessons was not deferred by the students. It was the teacher who guided the work of the class, and provided tasks and a brief introduction for students as well as a summary and evaluation of the students' results.

The requirements and curriculum also were very close to the Soviet model. In the 1980s a program of studies in the mathematics departments of the pedagogical institutes included courses of Real Analysis in One Variable and in Several Variables, Complex Analysis, Introduction to Functional Analysis and the Theory of Linear Operators, Measure Theory, Ordinary Differential Equations, Foundations of Mathematics, Abstract Algebra, Linear Algebra, Axiomatic Geometries, and Numerical Analysis and Programming. These courses proved difficult for most future teachers and later the most difficult courses were removed from the program. The scientific rigor of the curriculum began to decrease in the 1990s and seems not yet to have attained that envisioned in the 1980s.

The books and manuals in use very often were those translated from Russian and published by the Soviet publishing house MIR. Among the Soviet books translated were Kolmogorov and Fomin

(1978), Markushevich (1970), Kudriavtsev (1983), and Bugrov and Nikolsky (1988). Hundreds of mathematics publications of all sorts were prepared or translated for use at institutions of higher education and as supplementary and popular books for students and teachers. Books on mathematics education and on education in general also were translated from Russian or written with substantial reliance on Russian textbooks (Abdulina, 1984; Ballester *et al.*, 1992; Ballester and Arango, 1995).

Importantly, many foreign visiting professors were providing direct assistance by teaching in the mathematics faculties at universities. They helped to create the conditions for scientific research at an international level at Cuban universities and at the Cuban Academy of Sciences.

The best Cuban university graduates were sent abroad, mostly to the Soviet Union, East Germany, and Yugoslavia, to pursue scientific degrees. The first three to earn doctorates in mathematics abroad were one in East Germany in 1974, and two in 1976 from Moscow State University (Sánchez and Valdés, 2003). This number increased substantially in the following years until the end of the 1980s. Mathematics dissertations in these years were mostly supervised by foreign specialists, often from the Soviet Union. Thousands of other students were sent to pursue undergraduate studies in a variety of mathematics specialties including mathematics education, in a variety of institutions of these countries during the same period of time (Guadarrama, 2005).

6 The Situation Today

The maturity reached by Cuban mathematics education at all levels seems to be unprecedented in Latin America (with only Chile as a possible challenger). The mathematics and mathematics education research that has been conducted in Cuba since 1959 has received international prestige. In many instances it has been acknowledged that these accomplishments were the result of consistent support from the countries in the former Eastern Europe Socialist bloc, especially from the former Soviet Union. From the very beginning of this historical period, Cuba received economic support and benefited from international trade relationships among the socialist countries. The Ministries of Education and Higher Education counted on the assistance of Russian consultants in all content areas, including scholar organization, and curriculum development and implementation. Cuba multiplied this support by offering mathematical assistance to other Spanish-speaking nations with substantial success. Today many Cuban, Cuban-trained or Cuban-inspired mathematicians and educators serve several nations in Latin America effectively.

The collapse of the Soviet Union and the decline of its support to Cuba, that was so catastrophic for the Cuban economy, also proved to be disastrous for Cuban mathematics education. Many mathematical faculty members found themselves in the position of having to emigrate or search for another job due to the lack of funding or productive research collaboration. Difficult economic conditions at all educational levels substantially decreased the mathematical opportunities for Cuban youth. For example, the program of special instruction for the mathematically talented was eliminated or severely contracted. Opportunities for mathematics students from other nations to study mathematics in Cuban universities are severely limited. Financial support for participation or study abroad is now non-existent.

References

Abdulina, O. A. (1984). *La preparación pedagógica general del maestro en el sistema de instrucción superior pedagógica*. Moscú.

Ballester, S. *et al.* (1992). *Metodología de la enseñanza de la Matemática*. Editorial Pueblo y Educación. Tomo 1. Ciudad de La Habana.

Ballester, S., and Arango, C. (1995). *Cómo consolidar conocimientos matemáticos*. Havana, Editorial Academia.

Bugrov, Y., and S., Nikolski, S. M. (1988) *Matemáticas Superiores*, Vols. I, II, and III. Moscow: Editorial Mir.

Cruz-Taura, G. (2003). *Cuba Transition Project. Rehabilitating Education in Cuba: Assessment of Conditions and Policy Recommendations*. Institute for Cuban-American Studies, University of Miami, FL.

Guadarrama, P. (2005). Etapas principales de la Educación Superior en Cuba. *Revista Historia de la Educación Latinoamericana*, *7*, 51–74. Universidad Pedagógica y Tecnológica de Colombia, Tunja, Colombia.

Jungk, W. (1985). *Metodología de la enseñanza de la matemática*. Habana, Cuba: Pueblo y Educación.

Kolyagin, Iu. M., Lukankin, G. L., Oganesian, V. A., and Saninskii, V. Ya. (1975). *Metodika prepodavaniya matematiki v sredney shkole.* M.: Prosveschenie.
Kolmogorov, A. N., and Fomin, S. V. (1978). *Elementos de la Teoría de Funciones y del Análisis Funcional.* Moscow: Editorial Mir.
Kudriavtsev, L. D. (1983). *Curso de Análisis Matemático*, Vols. I and II. Moscow: Editorial Mir.
Markushevich, A. (1970). *Teoria de las Funciones Analíticas*, Vols. I and II. Moscow: Editorial Mir.
MINED (Ministry of Education). (1982). *Metodología de la enseñanza de la matemática.* Editorial Pueblo y Educación. Primera Reimpresión. Ciudad de la Habana.
MINED. (1993). *Estado actual y proyección de la enseñanza de la asignatura matemática.* Informe a la reunión de la Comisión Nacional de Matemática. Ciudad de la Habana. Octubre.
Sánchez, C., and Valdés, C. (2003). Bosquejo histórico de la actividad matemática en Cuba. *Boletín de la Sociedad Cubana de Matemática y Computación, 1*(1), 17–29.
Vogeli, B. R. (1997). *Special Secondary Schools for the Mathematically and Scientifically Talented. An International Panorama.* Teachers College, Columbia University.

10

Influences of Soviet Research in Mathematics Education

Jeremy Kilpatrick

University of Georgia, GA, USA

1 Introduction

Until the 1960s, most U.S. mathematics educators knew little or nothing of the body of research in the field that had been produced by Soviet researchers. In 1962, the American Psychological Association published an edited book (Bauer, 1962) in which a number of American psychologists reported on visits they had made to the Soviet Union in the summer of 1960. In Walter Reitman's (1962) chapter about studies on thinking and problem solving, one section dealt with what he called "pedagogically oriented research." Among the research studies Reitman discussed were studies by P. Ya. Galperin on teaching mathematical concepts to young children and by V. A. Krutetskii on mathematical ability. The book gave U.S. mathematics educators a glimpse of relevant activities in the Soviet Union, but there was little detail on the research itself.

Two events the following year brought Soviet research on mathematics education into somewhat clearer focus. The first was the publication in English of a collection of Soviet work in educational psychology (Simon and Simon, 1963). In that collection, there were two articles dealing directly with mathematics education, one by Z. I. Kalmykova (1955/1963) on arithmetic problem solving and the other by Krutetskii[1] (Krutetski, 1961/1963) on the thinking of pupils with little mathematical ability. Both topics were of great interest to U.S. mathematics teachers and researchers. The second event was the submission by Krutetskii (Krutetzky, 1964) of a brief paper on mathematical abilities to the Seventeenth International Congress of Psychology in Washington, DC. Krutetskii's efforts drew attention "not only because he appeared to be unique among Soviet psychologists in investigating individual differences in mathematical abilities, but also because the mathematical problems he used in his research were so varied and ingenious" (Kilpatrick and Wirszup, 1976, p. vii).

2 Educational Psychology in the Soviet Union and the United States

It soon became clear to American researchers that there was a rich trove of work in Soviet psychology that might throw light on problems of the teaching and learning of school mathematics. Whereas in the United States, educational psychology occupies only a small fraction of the field of psychology[2] — with little attention until recently to the psychology of learning and teaching specific subjects such as mathematics — in the Soviet Union, research in educational psychology was "the dominant area" (Brožek, 1966, p. 178), comprising in 1963, along with child psychology, almost 38% of all Soviet psychological publications. As Reitman (1962, p. 42) pointed out:

> Most Soviet psychology is organized under the Academy of Pedagogical Sciences of the Russian Soviet Federated Socialist Republic, and consequently reflects a much greater emphasis on working on important problems of educational practice than is the case in the United States.

---

[1] Krutetskii's name has been transliterated into English as Krutetski, Krutetzky, and Krutetskii. The last is preferred here.

[2] From 1957 to 1962, doctorates in educational psychology were less than 9% of those awarded in all fields of psychology in the United States (Harmon and Soldz, 1963, p. 13), and that percent has steadily declined.

Another striking difference between U.S. (and British) psychology and Soviet psychology was in their approach to studying the individual. In much Anglo-Saxon educational psychology, the learner's behavior is shaped in response to the environment, whereas as dialectical materialists, Soviet psychologists emphasized process over product, the dynamics of thought over the statics:

> To sum up those features of Soviet psychology which distinguish it most from its Anglo-Saxon counterpart, the former emphasizes the *active* part played by the subject (and especially the conscious human subject) in structuring his own environment and his own experience, in contrast to the traditional (though perhaps weakening) Anglo-Saxon insistence on a *passive* organism, in which associations are formed by the interplay of processes ... assuring successful *adaptation to* the environment (Gray, 1966, pp. 1–2).

These differences stimulated the curiosity of U.S. mathematics educators, particularly since their research attention was turning away from studies modeled on behavioristic laboratory experiments and toward more naturalistic studies of teaching and learning in mathematics classrooms. Much Soviet research in mathematics education took place in schools rather than laboratories, and it dealt with concepts from the school curriculum rather than artificial constructs — features that were especially attractive to U.S. researchers.

3 English Translations of Soviet Work

In response to the growing interest of U.S. mathematics educators in Soviet research that began during the 1960s, the series *Soviet Studies in the Psychology of Learning and Teaching Mathematics* was launched in 1969 as a joint project of the School Mathematics Study Group at Stanford University and the Survey of Recent East European Mathematical Literature at the University of Chicago. The series eventually included 14 volumes published between 1969 and 1975 by the School Mathematics Study Group. A second series of eight volumes, *Soviet Studies in Mathematics Education*, was published from 1990 to 1992; the first six volumes by the National Council of

Teachers of Mathematics, and the last two by the University of Chicago School Mathematics Project. Taken together, these books, along with the English translation of Krutetskii's (1968/1976) *The Psychology of Mathematical Abilities in Schoolchildren*, became the greatest source of information about Soviet research in mathematics education for English-speaking readers.[3]

3.1 *Soviet Instructional Psychology*

The first volume of the first series contained a survey by N. A. Menchinskaya (1967/1969) of 50 years of Soviet instructional psychology in which she showed how Soviet psychologists had over the years attempted simultaneously to formulate general principles of pedagogy and to develop specific instructional techniques that teachers could use. Characteristic of the Soviet approach was the way in which the psychologists doing the research in classrooms also played a major role in curriculum reform.

3.2 *A Measurement Approach to Number*

The final article in that same volume, by P. Ya Gal'perin and L. S. Georgiev (1960/1969), turned out to be especially influential in the United States. It elaborated Gal'perin's view that because measuring is a more elementary notion than counting, the teaching of measurement should precede the teaching of numerical concepts. An example of its influence on contemporary thinking can be seen in an article by Sophian (2004), in which she reports on a preschool mathematics instructional program based on the measurement approach used by Gal'perin and his coworkers, particularly V. V. Davydov (1966/1975), who had an important article in the *Soviet Studies in Psychology* series. Researchers at the University of Wisconsin subsequently invited Davydov (1982) to a landmark conference in 1979 on children's thinking about addition and subtraction.

---

[3]The journal of translations *Soviet Education*, which began publication in 1958, was also a widely used source.

4 Sociocultural Approaches

Much of the influence of Soviet psychology on U.S. research in mathematics education has come from the work of Lev Vygotsky (1896–1934), who had been Gal'perin's mentor. In fact, Gal'perin adapted Vygotsky's sociocultural approach to psychology and applied it to school instruction. When Vygotsky's book *Thought and Language* (1934/1962) was translated into English, U.S. mathematics educators began to learn of his emphasis on the role of language and culture in shaping human development, his explanation of the ways in which learners come to perceive their world, and his views on consciousness and how the mind reflects on itself. These ideas provided an intriguing contrast to those of Jean Piaget, whose books were being translated and becoming popular in the United States at the same time. The later publication, *Mind in Society*, was undoubtedly even more influential in spreading Vygotsky's (1978) ideas. But it took some time for those ideas to percolate. Stephen Lerman (2000, p. 25), documenting early references to Vygotsky in the literature, claims that it was not until the late 1980s that the mainstream mathematics education community took note of Vygotsky's work. Ellice Forman (2003) has recently documented the influence of sociocultural ideas from Vygotsky and others on efforts to reform the U.S. school mathematics curriculum.

Researchers in mathematics education have been especially attracted by Vygotsky's (1978, p. 86) best known idea: the *zone of proximal development* (ZPD), defined as "*the distance between the actual developmental level as determined by independent problem solving and the level of potential development as determined through problem solving under adult guidance, or in collaboration with more capable peers.*" That is, the ZPD is the difference between assisted and unassisted performance. Researchers in mathematics education have been interested in how novices gain expertise in solving problems by being assisted in various ways (through hints, by working with others, etc.). In Vygotsky's theory, the learner moves from regulation by others to self-regulation by being given assistance in the ZPD. As an example of researchers' attention to the ZPD, Anderson Norton and Beatriz D'Ambrosio (2008) used the ZPD construct to analyze the mathematical constructs that a teacher fostered in working with two sixth graders over the course of a semester.

5 Teaching Experiments

Norton and D'Ambrosio (2008) termed their research study a "teaching experiment." That term has been borrowed by U.S. researchers from Soviet work, but it has come to have a somewhat different meaning. Menchinskaya (1967/1969, p. 6) describes the Soviet teaching experiment as follows:

> The researcher attempts to develop essential knowledge, abilities, skills, and modes of activity in the child during the experiment in order to reveal the child's psychological traits in the making, in their dynamic state. The teaching experiment requires the introduction of preliminary stages of study, in which the initial information, abilities and skills needed to master the new material are ascertained and "levelled" (arranged in a hierarchy).
>
> And, finally, in recent years a new form of teaching experiment has begun to be used on a broad scale: The experiment is done with entire classes over a number of years, with changes not only in teaching methods but in the curriculum as well.

Some U.S. researchers in mathematics education attempted to follow the Soviet model (Kantowski, 1978; Lester, 1985), but others have modified it. They do not necessarily arrange the material to be taught in a hierarchy, and they often work with individuals, pairs, or small groups of students rather than adopting the "new form" in which the Soviet researcher worked with an entire class (Cobb, 2000; Norton and D'Ambrosio, 2008). An entire section of the Kelly and Lesh (2000) handbook — five chapters — is devoted to teaching experiments, and those chapters demonstrate the different ways in which U.S. researchers have interpreted the Soviet idea.

6 Mathematical Abilities

The pioneering work of Krutetskii is represented in Volume 2 of the *Soviet Studies in Psychology* series, Volume 8 of the *Soviet Studies in Mathematics Education*, and the 1976 book. As noted above, his work is unique. In part, that uniqueness arises from the political climate in which he worked. Anyone attempting to study individual differences in mathematical abilities had to contend with the 1936 ban on mental tests by the Central Committee of the Communist Party. The official Soviet view was that although achievement tests could still be used to measure progress in learning, other tests were prohibited. Any test provided an index of current status only. It gave no information on a pupil's potential level of performance or on the processes the pupil used to respond to the test items. Testing encouraged the labeling of pupils and the setting of norms for their performance. It hindered the development of effective instructional procedures. Consequently, Krutetskii and his students had not only to avoid claiming that some pupils were necessarily more able than others but also to develop alternative methods for assessing their mathematical abilities.

Because Krutetskii had to avoid testing and the use of factor analysis to detect components of mathematical ability, he was freed to use a wide range of tasks that went deeper in to how students approached mathematical problems and what they were thinking as they solved them. Assessment was done through a series of interviews in which pupils were asked to think aloud while solving a variety of challenging problems. Hints and suggestions were given, and problems were sometimes changed in an effort to find out how the pupil responded to varied conditions. Many of the mathematical problems were quite clever, and it is no surprise, therefore, that they have been copied and adapted by researchers in mathematics education. As just one example, from Canada, see Kaizer and Shore (1995).

7 Conclusion

Although the Soviet research literature has come to English-speaking researchers in mathematics education almost entirely through translations, many of which have not been easy for them to learn about let alone procure, it has nonetheless stimulated a great variety of studies. U.S. researchers, in particular, were ready to move away from examining individual students' learning, and the Soviet approach allowed them to see the advantages of looking at learning as it takes place in school mathematics classrooms. Soviet research in mathematics education widened the horizons of many mathematics educators around the world, showing them what was possible when one approached learning, teaching, thinking, and research itself as sociocultural phenomena.

References

Bauer, R. A. (Ed.). (1962). *Some Views on Soviet Psychology.* Washington, DC: American Psychological Association.

Brožek, J. (1966). Contemporary Soviet psychology. In N. O'Conner (Ed.), *Present-Day Russian Psychology* (pp. 178–198). Oxford: Pergamon.

Cobb, P. (2000). Conducting teaching experiments in collaboration with teachers. In A. E. Kelly and R. A. Lesh (Eds.), *Handbook of Research Design in Mathematics and Science Education* (pp. 307–333). Mahwah, NJ: Erlbaum.

Davydov, V. V. (1975). The psychological characteristics of the "prenumerical" period of mathematics instruction. In L. Steffe (Ed.), *Children's Capacity for Learning Mathematics.* Soviet Studies in the Psychology of Learning and Teaching Mathematics, Vol. 7 (pp. 109–205). Stanford, CA: School Mathematics Study Group (Original work published 1966).

Davydov, V. V. (1982). The psychological characteristics of the formation of elementary mathematical operations in children. In T. P. Carpenter, J. M. Moser, and T. A. Romberg (Eds.), *Addition and Subtraction: A Cognitive Perspective* (pp. 224–238). Hillsdale, NJ: Erlbaum.

Forman, E. A. (2003). A sociocultural approach to mathematics reform: Speaking, inscribing, and doing mathematics within communities of practice. In J. Kilpatrick, W. G. Martin, and D. Schifter (Eds.), *A Research Companion to Principles and Standards for School Mathematics* (pp. 333–352). Reston, VA: National Council of Teachers of Mathematics.

Gal'perin, P. Ya., and Georgiev, L. S. (1969). The formation of elementary mathematical notions. In J. Kilpatrick and I. Wirszup (Eds.), *The Learning of Mathematical Concepts.* Soviet Studies in the Psychology of Learning and Teaching Mathematics, Vol. 1 (pp. 189–216). Stanford, CA: School Mathematics Study Group (Original work published 1960).

Gray, J. A. (1966). Attention, consciousness and voluntary control of behaviour in Soviet psychology: Philosophical roots and research branches. In N. O'Conner (Ed.), *Present-Day Russian Psychology* (pp. 1–38). Oxford: Pergamon.

Harmon, H. P., and Soldz, H. (1963). *Doctorate Production in United States Universities 1920–1962, with Baccalaureate Origins of Doctorates in Sciences, Arts and Professions.* Washington, DC: National Academy of Sciences (ERIC Document Reproduction Service No. ED054712).

Kaizer, C., and Shore, B. M. (1995). Strategy flexibility in more and less competent students on mathematical word problems. *Creativity Research Journal, 8*, 77–82.
Kantowski, M. G. (1978). The teaching experiment and Soviet studies of problem solving. In L. L. Hatfield and D. A. Bradbard (Eds.), *Mathematical Problem Solving: Papers from a Research Workshop* (pp. 43–52). Columbus, OH: ERIC Clearinghouse for Science, Mathematics, and Environmental Education.
Kalmykova, Z. I. (1963). Psychological prerequisites for increasing the effectiveness of learning in problem solving in arithmetic. In B. Simon and J. Simon (Eds.), *Educational Psychology in the USSR* (pp. 180–191). Stanford, CA: Stanford University Press (Original work published 1955).
Kelly, A. E., and Lesh, R. A. (Eds.). (2000). *Handbook of Research Design in Mathematics and Science Education*. Mahwah, NJ: Erlbaum.
Kilpatrick, J., and Wirszup, I. (1976). Editors' preface. In V. A. Krutetskii, *The Psychology of Mathematical Abilities in Schoolchildren* (J. Teller, Trans., pp. vii–viii). Chicago: University of Chicago Press.
Krutetski, V. A. (1963). Some characteristics of the thinking of pupils with little capacity for mathematics. In B. Simon and J. Simon (Eds.). *Educational Psychology in the USSR* (pp. 214–232). Stanford, CA: Stanford University Press (Original work published 1961).
Krutetskii, V. A. (1976). *The Psychology of Mathematical Abilities in Schoolchildren* (J. Teller, Trans.). Chicago: University of Chicago Press (Original work published 1968).
Krutetzky, V. A. (1964). A study of structure, ways of development and diagnosis of mathematical abilities. *Acta Psychologica, 23*, 314–315.
Lerman, S. (2000). The social turn in mathematics education research. In J. Boaler (Ed.), *Multiple Perspectives on Mathematics Teaching and Learning* (pp. 19–44). Westport, CT: Ablex.
Lester, F. K., Jr. (1985). Methodological considerations in research on mathematical problem-solving instruction. In E. A. Silver (Ed.), *Teaching and Learning Mathematical Problem Solving: Multiple Research Perspectives* (pp. 41–69). Hillsdale, NJ: Erlbaum.
Menchinskaya, N. A. (1969). Fifty years of Soviet instructional psychology. In J. Kilpatrick and I. Wirszup (Eds.), *The Learning of Mathematical Concepts*. Soviet Studies in the Psychology of Learning and Teaching Mathematics, Vol. 1 (pp. 5–27). Stanford, CA: School Mathematics Study Group (Original work published 1967).
Norton, A., and D'Ambrosio, B. S. (2008). ZPC and ZPD: Zones of teaching and learning. *Journal for Research in Mathematics Education, 39*, 220–246.
Reitman, W. R. (1962). Some Soviet investigations of thinking, problem solving, and related areas. In R. A. Bauer (Ed.), *Some Views on Soviet Psychology* (pp. 29–61). Washington, DC: American Psychological Association.
Simon, B., and Simon, J. (Eds.). (1963). *Educational Psychology in the USSR*. Stanford, CA: Stanford University Press.
Sophian, C. (2004). Mathematics for the future: Developing a Head Start curriculum to support mathematics learning. *Early Childhood Research Quarterly, 19*, 59–81.

Vygotsky, L. S. (1962). *Thought and Language* (E. Hanfmann and G. Vakar, Eds.). Cambridge, MA: MIT Press (Original work published 1934).
Vygotsky, L. S. (1978). *Mind in Society: The Development of Higher Psychological Processes* (M. Cole, V. John-Steiner, S. Scribner, and E. Souberman, Trans.). Cambridge, MA: Harvard University Press.

Notes on Contributors

Alexander Abramov graduated from the Department of Mechanics and Mathematics at Moscow State University and defended a candidate's dissertation (Ph.D.) under Andrey Kolmogorov. Concurrently, he taught mathematics at MSU's Physics-Mathematics School. Subsequently, Abramov worked in the publishing house *Prosveshchenie* and at the Scientific Research Institute on Educational Content and Methods under the aegis of the USSR Academy of Pedagogical Science. From 1988 until 1990, he was part of the "School" task force, which prepared documents for school reforms. In 1990–1991, he served as advisor to the Russian minister of education. Abramov was the founder and director of the Moscow Institute of Educational Systems (MIROS). In 1992, he was elected a corresponding member of the Russian Academy of Education. Abramov is the author of over 300 articles on mathematics teaching methodology and general problems of education. He won renown for his polemical articles and appearances on radio and television. Currently, Abramov's main focus is on preparing a multivolume edition of the works of Andrey Kolmogorov devoted to the problems of mathematics education.

Orlando B. Alonso is a Cuban-American mathematics teacher who immigrated to the United States in 1996. He is a former teacher at the Ernesto Guevara Vocational School and served as Associate Professor at The Felix Varela Higher Pedagogical Institute in Santa Clara, Cuba. Mr. Alonso has furthered his education in New York City, pursuing two doctorate degrees in the fields of Mathematics Education and Mathematics at The Teachers College of Columbia University and The City University of New York, respectively. Mr. Alonso has taught mathematics at Louis Brandeis High School and has served as an Instructor within The City University of New York. Currently, he works in the Division of Education at Mount Saint Mary College.

Mark Bashmakov defended both his candidate's and doctoral dissertations in mathematics at Leningrad State University. In 1963, he began teaching at the university's mathematics-mechanics department. Beginning in 1977 and for the next 15 years, he was chair of the mathematics department at the Leningrad Electrotechnical Institute. In 1991, Bashmakov founded the Institute of Productive Teaching, whose director he remains to this day. In 1993, he was elected a member of the Russian Academy of Education. Bashmakov has organized many mathematics Olympiads, clubs, and schools; he was an active participant in the creation of the magazine *Kvant*. Bashmakov has participated in and organized many Russian and international projects. He is the author of numerous scholarly articles on mathematics, mathematics education, and the general problems of education, as well as popular-scientific books and articles, teaching manuals, and textbooks. Bashmakov was also a professional mountain climber and a recipient of the Snow Leopard Trophy.

Dmitri Fomin was born in St. Petersburg (then Leningrad), Russia. Fomin attended the famous 45th Specialized (Science and Math) High School affiliated with Leningrad State University, on whose team he participated in two All-Russia and then All-Union Mathematics Olympiads. Fomin graduated with honors in 1986 from the Faculty of Mathematics and Mechanics at Leningrad State. He majored in Geometry and Topology, to which his post-graduate studies in 1987–89 were also devoted. While working as an assistant professor and later as senior lecturer at the Chair of Higher Geometry and Topology at St. Petersburg State University, he was actively involved in conducting Olympiads of different levels and teaching mathematics circles. Fomin moved to the United States in 1994 and currently works as a Technology Fellow at the Geometric Applications Department of Research and Development at the Parametric Technology Company in Needham, Mass. Fomin has authored more than 30 articles and several books on topology, algebra, combinatorics, mathematics education and competitions. He has given lectures on mathematics education and competitions in Russia, Australia, and the United States.

Katalin Fried was born in Budapest, Hungary. Her father is a well known Hungarian mathematician and her mother is a mathematics teacher who worked as the editor-in-chief of *KöMaL*, a well-known Hungarian journal of mathematics and physics for high school students. Katalin Fried received her Ph.D. in mathematics in 1999. Since 1983 until the present she has been teaching mathematics to prospective teachers at Eötvös Loránd University, Budapest. Currently, her research interests are mainly focused on mathematics teacher education. She has been a presenter at a number of conferences on mathematics education and her publications include books on teaching mathematics and several textbooks. She is married and has two children.

Alexander Karp is an associate professor of mathematics education at Teachers College, Columbia University. He received his Ph.D. in mathematics education from Herzen Pedagogical University in St. Petersburg, Russia, and also holds a degree from the same university in history and education. For many years, Karp worked as a teacher in a school for mathematically gifted in St. Petersburg and as a teacher educator. Currently, his scholarly interests span several areas, including gifted education, mathematics teacher education, the theory of mathematical problem solving, and the history of mathematics education. He is the managing editor of the *International Journal for the History of Mathematics Education* and the author of over 90 publications, including over 20 books.

Jeremy Kilpatrick is Regents Professor of Mathematics Education at the University of Georgia. He has also taught at European and Latin American universities, receiving four Fulbright awards. He holds an honorary doctorate from the University of Gothenburg, is a National Associate of the National Academy of Sciences and a Fellow of the American Educational Research Association, and has a Lifetime Achievement Award from the National Council of Teachers of Mathematics and the 2007 Felix Klein Medal from the International Commission on Mathematical Instruction. His research interests include proficiency in mathematics teaching, curriculum change and its history, assessment, and the history of research in mathematics education.

Antoni Pardała, D.Sc., Ph.D., is an Associate Professor at the Department of Mathematics in the Faculty of Mathematics and Applied Physics at the Rzeszów University of Technology. He received his Ph.D. in Kraków and his D.Sc. (post-doctoral degree) in Moscow (both in Mathematics Education). Among his advisers were A. Z. Krygowska, G. V. Dorofeev and I. F. Sharygin. His research interests include teaching and learning mathematics both at secondary and university levels as well as assessment and evaluation in mathematics education. Dr. Pardała is involved in studying both theoretical and practical aspects of mathematics education (in particular, the organization of gifted education). He has authored 56 research articles, which were published in Polish, English and Russian, including two monographs.

Tatiana Polyakova is a professor and head of the Department of Geometry and Mathematics Education at the Pedagogical Institute, Southern Federal University (Russia). Her Ph.D. dissertation addressed the difficulties of beginning teachers in mathematics teaching, and her post-doctorate degree (habilitatus) was devoted to the use of the history of mathematics education in teacher education. She has more than 150 publications including a book based on her Ph.D. dissertation translated into German and a number of books on the history of mathematics education in Russia.

Mark Saul got his B.A. from Columbia University and Ph.D. from New York University. He spent 35 years in and around New York, teaching mathematics in classrooms from grades 3 through 12. He is a 1984 recipient of the Presidential Awards for Excellence in Mathematics and Science Teaching. He has served as President of the American Regions Mathematics League, mathematics field editor of *Quantum* (the English-language version of the Russian journal *Kvant*), board member of the National Council of Teachers of Mathematics, and member of the Mathematical Sciences Education Board for the National Research Council. More recently, he served as a program director for the National Science Foundation. Internationally, he initiated a student exchange program between Russian and American students, has given talks, done consulting work, and led workshops in 20 countries. His publications include several problem books, an elementary text on trigonometry, co-authored with I. M. Gelfand, and a translation and reader's companion for Jacques Hadamard's "Lessons in Elementary Geometry." He is currently a Senior Scholar for the John Templeton Foundation.

Jean Schmittau holds a Ph.D. from Cornell University in Educational Psychology, Cognitive Development, and Mathematics, and is a professor at the State University of New York at Binghamton. She is the editor of *Investigations in Mathematics Learning* and directs the Teacher Leader Quality Partnership, a project to improve middle school mathematics instruction in upstate New York school districts. Her interest in Vygotskian psychology led to research in Russia with students who had been instructed with V. V. Davydov's mathematics curriculum, and subsequently to editing its English translation and researching its first implementation in a U.S. school setting.

Alexey Sossinsky was born in Paris of Russian émigré parents, has a French high school education, a B.S. degree from New York University, and an M.S. and Ph.D. from Moscow State University. A research mathematician, he has always had a strong interest in mathematics education and in the popularization of mathematics. He began his mathematics career as an assistant and then associate professor at Moscow State University and at that time also taught in Kolmogorov's physics and mathematics boarding school and was active in the Olympiad movement. Forced to leave MSU for political reasons in 1974, he worked for 13 years in the popular science magazine *Kvant.* At present, he is a professor at and the vice president of the Independent University of Moscow. He was an invited speaker and a panel coordinator at the International Congress of Mathematical Education (Copenhagen, 2004) and a panelist at the round table on the popularization of mathematics at the International Congress of Mathematicians (Madrid, 2006). Sossinsky is the author of over 50 research articles, several mathematical monographs, and popular science books, including a book on knot theory that has been translated into six languages.

Natalya Stefanova was born in the Russian Far East. In 1973, she graduated from Herzen Leningrad State Institute (now Herzen

Pedagogical University of Russia), specializing in mathematics and obtaining a license to teach mathematics at the secondary school level. Stefanova holds two academic degrees in the field of mathematics education methodology (a candidate's degree and a doctoral degree) and is a professor at Herzen Pedagogical University. For 17 years, she served as dean of the mathematics department at Herzen Pedagogical University. Currently, she is the chair of the department of mathematics education methodology. Stefanova has received the honorary title of Distinguished Worker of Russian Higher Education. She is the author of over 80 scholarly articles, has taken part in numerous conferences in Russia and abroad, and has served as research advisor for 15 candidate's degree (Ph.D.) students.

Bruce R. Vogeli is the Clifford Brewster Upton Professor and Director of the Program in Mathematics at Teachers College, Columbia University. Before joining the Columbia University faculty in 1965, Professor Vogeli served as a Visiting Professor at the Lenin Pedagogical Institute in Moscow. He also lectured at the Herzen Pedagogical Institute in Leningrad and worked as a "curriculum specialist" in the Mathematics Section of the Academy of Pedagogical Science in Moscow. Professor Vogeli is the author of *Soviet Schools for the Mathematically Talented* and *Special Mathematics Schools — An International Panorama* as well as more than 200 books for students and teachers.

Name Index

Subject Index